Francis W. Smith

The Natural Waters of Harrogate

Chemically, therapeutically and clinically considered with reference to

their application by drinking and bathing: by the light of fresh analysis, and

by examination of the blood

Francis W. Smith

The Natural Waters of Harrogate
*Chemically, therapeutically and clinically considered with reference to their
application by drinking and bathing: by the light of fresh analysis, and by
examination of the blood*

ISBN/EAN: 9783337251864

Printed in Europe, USA, Canada, Australia, Japan

Cover: Foto ©berggeist007 / pixelio.de

More available books at **www.hansebooks.com**

THE

NATURAL WATERS OF
HARROGATE

CHEMICALLY, THERAPEUTICALLY, & CLINICALLY
CONSIDERED, WITH REFERENCE TO THEIR
APPLICATION BY DRINKING & BATHING;
BY THE LIGHT OF FRESH ANALYSIS,
& BY EXAMINATION OF THE BLOOD.

BY

FRANCIS WILLIAM SMITH, M.D.,

AND BACHELOR OF SURGERY,
AUTHOR OF "THE SALINE WATERS OF LEAMINGTON," AND "HEALTH
RESORTS OF THE WORLD," ETC., ETC.

LONDON:

DAWBARN AND WARD, LIMITED,
6 FARRINGDON AVENUE. E.C.

1899.

Dawbarn & Ward
Limited

I DEDICATE THIS WORK

TO

THE DOCTORS OF THIS COUNTRY

FOR

THE GOOD OF THEIR

PATIENTS

AND

THE GLORY OF HARROGATE.

EXPLANATORY PREFACE.

I AM writing this little work on exactly the same lines as I did a similar one on the Mineral Waters of Leamington.

I thank Mr Symons, F.R.S., the author of *British Rainfall*, for lending me his books, so as to enable me to compile a table of statistics for the last ten years; also Mr Turner-Taylor, the Town Clerk of Harrogate, for information with reference to the death-rate. I likewise owe a debt of gratitude to Mr G. Paul, of Harrogate, for his able assistance in the geological and climatological part of the work.

I have arranged a table of the mineral waters of Harrogate for comparison with those of Germany and others of this country, from works of Braun and Martindale, and for the assistance in compiling this I thank Mr Eynon, chemist, Harrogate.

In the context, wherever specific views are given upon any subject, I have inscribed the name of the authority for the same.

Finally, my thanks are due to that distinguished physician, Dr Mitchell Bruce, an old fellow-student and a life-long friend, for allowing me to pick some crumbs from his masterly work on Materia Medica and Therapeutics.

It is just possible that by the time this work is in print, what is called "an official analysis" of the Harrogate waters may be published by the Corporation.

When that is done, I will instruct my publishers to send a copy of the same to all purchasers of this work.

The reader, however, may rely upon my new analysis, which he will find later on in these pages, being, in every respect, thoroughly trustworthy and accurate.

Assyst,
        Harrogate,
                1899.

PREFACE.

THERE is an old saying, that "a good tale cannot be too often told." And so I think now. The theme of the efficacy of the Harrogate Waters can never be exhausted, any more than the running of its healing streams. The springs will continue anon to well up, and their babbling brooks flow on to the sea.

I make, therefore, no apology for adding one more work upon the subject of their natural virtues. For

> "One doctor only, like the sculler, tries,
> The patient struggles and at last he dies ;
> But two physicians, like a pair of oars,
> Will land him gently on the Stygian shores."

I have had the waters from six of the Wells mostly in use (out of the many) subjected to a fresh analysis by chemists of the highest reputation at Charing Cross Hospital. I have applied the treatment by these waters to certain diseases, as indicated by examination of the blood and other methods, and I have put to the test all the latest appliances in balneo-therapeutics, as used at Harrogate. I have endeavoured to bring the whole "up to date," and to make this work a *standard* upon the subject—for the time being—on all points fair, without being over-drawn.

It is not intended to make this a general guide book; at the same time, it may be well for the Medical Profession to know that Harrogate, in addition to its mineral waters, has many natural charms in itself and in its immediate neighbourhood.

It is true that it cannot boast of the antiquity of Bath, with its Roman remains and baths; nor, like Royat and Dax, and half-a-dozen other Continental Spas, claim that Julius Cæsar was cured of his gout by drinking its waters and frequenting its baths; but it is nevertheless true that the healing properties of the waters of Harrogate were known and appreciated for many generations.

On the general treatment of ailments for which sufferers come to Harrogate, I have brought to bear the latest researches of examination of the blood, so as to arrive at an accurate diagnosis. And as a sequence and a consequence, I have been able to direct certain lines of treatment to suit individual cases with marked success, and this is all the more satisfactory, because they are based on scientific, physiological, and pathological principles. This will all be explained in the chapters which are to follow.

In writing this work on the Sulphur and other Mineral Waters of Harrogate, I will avoid verbosity, and describe them and the ailments they are likely to cure or ameliorate as tersely as possible. I have no intention to go into a lengthened comparison of Continental and other Mineral Waters, although I may touch upon them in a cursory way.

In placing Harrogate before the Profession in its medical bearings, I will describe it geographically; meteorologically; and, further, its drainage; mortality; seats of education; its geological features and mineral waters; and their value in their chemical, therapeutical, and clinical aspects.

In dealing with the Harrogate Mineral Waters and their efficacy in the treatment of disease, there are certain things which cannot as yet be explained.

Medicine, unfortunately, is not, like mathematics, a fixed and certain science: constitutions differ, and so does the effect of certain drugs upon individuals; for although the composition of all mineral waters may be well known, the mode of properly assimilating their component parts can only be done in Nature's laboratory.

We have, in the waters of many of our Wells, simple prescriptions, physiologically correct as to their application to certain diseases, but whose constituents are so blended together as to baffle the most skilful chemist to imitate successfully.

CONTENTS.

THE NATURAL WATERS OF HARROGATE.

INTRODUCTION.

IT is impossible to exaggerate the importance of embracing favourable opportunities which present themselves to individuals and communities. Opportunity makes the general, and opportunity makes the millionaire.

"There is a tide in the affairs of men which, taken at the flood, leads on to fortune." There can be no question that the present is Harrogate's golden opportunity. It has put its house in order and is ready to receive its guests.

English Spas are fast rising in public favour. Many physicians in London and the provinces, instead of exposing their patients to the annoyance and fatigue of foreign travel, are very prudently recommending them to use our own mineral springs, and it cannot be too often repeated that there is now no need for invalids to run the risk, or undergo the in-

conveniences and discomforts of a journey to the
Continent, in order to drink mineral waters. We
have as good sulphur, iron, and magnesia springs in
Harrogate as there are anywhere.

In our "temple of health," those who are sick may
rest assured of finding relief, and often a cure, for
many of " the ills that flesh is heir to," and it will be
well if patients, who are wearied and worried with the
rush to Baden-Baden, Homburg, and Kissingen, and
the fleecing and flaying experienced in these places,
would take this to heart.

About the year 1591 the first mineral spring of
Harrogate is said to have been discovered by Sir Wm.
Slingsby of Knaresborough.   It is known as the Tewitt
Well. He being a gallant and a roving knight, and
having travelled much in Germany and other parts,
had acquired a knowledge of mineral waters there. It
is said he made the discovery of the Tewitt Well
one day while hunting in the forest, which extended
for thousands of acres around Harrogate. In any
case, he appreciated its medicinal virtues, and took
regular courses of its waters, from year to year, up to
the time of his death in 1634.   These waters belong
to the Iron group—as do those from St John's Well—
discovered in 1631 by Dr Stanhope.

The exact date of the discovery of the Old Sulphur
Well is not known, but it must have been somewhere
between 1600 and 1626, for it was then that the first
medical work on the Harrogate waters was written
by Dr Dean.

Dr Stanhope was, at that time, in the height of his glory, and the fame of the healing springs gradually spread.

As years went on, the forest was cleared, roads were made, and cottages built — afterwards to be succeeded by larger houses and inns.

Strangers and visitors from the immediate neighbourhood, and even from a distance, came and drank the waters, chiefly as an antidote to scrofulous affections.

Many and varied are the accounts given of the modes of applying these waters—only to be exceeded by the wonderful cures wrought thereby. Not content with sounding their praise in prose, the goddess of poetry was invoked, and rhymes by the yard were spun and sung at the tap of the Queen's Head Inn.

And so the village grew and prospered—fresh mineral wells were discovered—now amounting to between eighty and ninety (all more or less different), hotels were built, streets laid out, baths erected; and finally a Charter of Incorporation was granted, and Harrogate blossomed into a Borough, with its Mayor, its mace, and Corporation.

GEOGRAPHICALLY CONSIDERED.

Harrogate, with its population of nearly 20,000 inhabitants, is situated in the West Riding of Yorkshire, midway between the German Ocean and the

Irish Channel, and the line, of 54 degrees latitude, on
which Harrogate stands, runs through Bridlington
Bay on the east coast, and Morecambe Bay on the west.
The highest parts of Harrogate are 600 feet above the
sea level, and Low Harrogate about 350 feet. It is
199 miles from London, and 212 from Edinburgh, each
of which places can be reached in comfort by train,
in between four and five hours without a change.

Of late years much has been done in the way of
building, and laying out the town in well-made streets,
shops and private residences, to say nothing of the
Royal Baths, and the addition of many excellent
hotels.

In 1779, the STRAY was set apart for the use of
the inhabitants " for ever," and never can it be built
upon. It is a fine open grass space of 200 acres, and
no matter to what extent Harrogate may increase, it
will always be looked upon as the lungs and heart of
the borough. (See the Map.)

The country around is varied in scenery—combining moorland in one part, great fertility in another.
It is undulating, and in parts wooded, and abounds
with objects of the greatest historical interest—Bolton
Abbey—with all its associations of "the Olden Time";
Fountains Abbey—the finest ecclesiastical ruin in the
country; Ripon and its cathedral—rich in its history
of raids by the Northern Scot; Knaresborough Castle,
at one time owned by John o'Gaunt, fourth son of
Edward the Third; and other places of minor historical
interest, such as Ribston Hall, with its world-famed

ROYAL BATHS, HARROGATE.

pater pippin tree, Ripley Castle, and stately Harewood.

Less than a mile from Harrogate is Harlow Tower, on Harlow Moor. Its summit is 670 feet above the sea, and around its base, for many acres, the moor is covered with lofty pines, rose-hued heather, and many a blooming wild shrub. In the summer, and even at other seasons of the year, the air is scented with fragrant perfumes, the ear is lulled by the humming bee, and the mind drawn upwards by the song of the lark.

From the tower, on a clear day, a splendid panoramic view is met with. Away out there is seen the bold outline of the Yorkshire Wolds, the hills and crags of rocky Hambleton, the wooded clump of Studley Royal, the stately pile of York Minster, the peaks of Derbyshire, the Cathedral of Lincoln, the lovely valleys of the Wharfe and Nidd, many an ancient mansion and scores of heavenward pointing church spires.

Need I say that, to the weary and ill-stricken visitor, these enchanting surroundings breathe on him their balm.

METEOROLOGICALLY CONSIDERED, AS TO DRYNESS.

Taking a period of ten years, and comparing the following inland watering-places, it will be seen that, on the whole, Harrogate stands well. If the reader will glance at the table of comparison, he will see

B

that it is so. Its dryness, from this point of view, is therefore established. The calculations are taken from Symons' *British Rainfall*, giving the average annual rainfall and number of days on which $\frac{1}{100}$ in. or more of rain fell.

METEOROLOGICAL TABLE.

Rainfall and "Rainy Days," 1888 to 1897 inclusive. From SYMONS' *British Rainfall*.

	10 Years' Average Rainfall.	10 Years' Average Rainy Days.	Observers.
	Inches.	No.	
Leamington, . .	21 86	160	Burnitt, on the Parade.
Cheltenham, . .	25·10	172·3	Kaye, Smelt, Tyner, and others.
Great Malvern, .	25·41	153·8	Mander, Munn, Crump.
Tunbridge Wells,	28 36	168	Weston, Winton.
HARROGATE, .	28·78	193	Wilson, Farrah, Dixon, Gledhill, Paul.
Matlock Bath, .	29·51	Not given.	Chadwick.
Bath, . .	31·46	157·6	Weston, Gilby, Institute.
Clifton, ...	32·66	176·5	Burder, Rintoul, Bridge, Sturge.
Ilkley, ...	34·81	168	Worfolk, Richardson.
Buxton, . . .	48·6	214·2	Thresh, Beck, R. Met. Soc., and others.
Means, .	30·65	173·7	

Harrogate stands on a plateau 400 to 500 feet above the sea. This does not necessarily detract from its dryness.

Other considerations have to be taken into account, and Harrogate, fortunately, is happy in these. The rainfall in any locality is determined by its mountainous surroundings.

Hilly outposts of any place materially modify the rainfall, and affect the dryness of it, especially if their ridges run at right angles to the direction of the winds which generally bring the rain.

The Pennine Range, with its various spurs, on which stand Skipton, Ilkley, and Otley, forms an outer rampart to check the moisture-laden winds coming from the Irish Sea on the west. These winds, charged with invisible moisture, strike the spurs of the Pennine Range in their approach to Harrogate, more or less at right angles, and in doing so, they *rise into a region of lower temperature, in proportion to the velocity with which they travel, and the contained moisture is condensed and precipitated as rain on the tops of the hills and on the immediate ground of the lee side.*

The air gets *warmer by the act of falling on the lee side,* and is *necessarily drier.* The south, south-west, and west winds are the rain-carrying ones in these parts, and having to pass over these mountain spurs, they thus exhaust their humidity to a great extent before they reach Harrogate. The rivers Wharfe and Aire are in this way fed by the expended efforts of moisture-laden winds passing over their watersheds, and Harrogate *is left drier*. Witness the respective rainfalls of Ilkley and Harrogate.

Absence of rivers and the geological formation of the subsoil also conduce to the general dryness of the atmosphere in Harrogate.

For the most part, the millstone grit overspreads Harrogate generally, but the Yoredale series prevails in Low Harrogate, and these are both of an absorbent nature in themselves, especially millstone grit, besides which the rocks in many places present their truncated edges upwards, and so act as conduits to lead the rainfall under the surface. And this they do in a *marvellously short time*.

TEMPERATURE.

I have had some difficulty in making the undermentioned calculations, but considered generally, the table represents pretty accurately the mean temperature for ten recent years :—

		Fahr.
Buxton	. . . .	45°·20
Leamington	. . .	48°·0
Cheltenham	. . .	48°·3
Clifton	. . . .	48°·7
HARROGATE	. . .	49°·0
Torquay	. . . .	49°·8
Bath	. . . .	50°·3
Bournemouth	. . .	50°·3
Llandudno	. . . .	50°·5

In considering the various factors which conduce
to the temperature of Harrogate, several items must
be taken into consideration, jointly and severally.
It is not fair, in reckoning the mean temperature of
a health resort, to gauge it by one part of the year—
that is, the summer (say from May till November),
and the winter (from November till May). In the
summer months, Harrogate requires no laudation as
to dryness and sunshine; in the winter, it has more
than its share. High altitudes, in winter, with dryness and sunshine, are all to the front in the present
time, and Harrogate stands well. To be short,
precise, and explanatory—"the mean temperature"
may be the same in two places, and yet the thermal
conditions in one place may be intolerable and un-
suitable for delicate persons, and the other may be
just as genial. The *maximum* temperatures of Harrogate are lower than in some Health Resorts in the
South, but the readings of the *minimum* thermometer
range higher generally, resulting in a coldness, which
is more bearable—thus giving a temperature which is
as tolerable as most of those places which claim to
have warmer climates.

THE CLIMATE MEDICALLY CONSIDERED.

From the foregoing remarks, it is proved that
the air of Harrogate is particularly dry and not
so cold as people imagine. It therefore follows

that it may be truly said to be a suitable place for residence all the year round. Many people, in the winter, can better bear and feel invigorated by a dry cold, than when they shiver and feel chilled by a "damp cold." And this does not apply only to the strong ; but weak patients also, when serious disease does not exist, bear the bracing air and enjoy it. Of course, one hears of people being dissuaded from making Harrogate a place of permanent residence. But this is by those who are not acquainted with the facts.

Persons who have suffered from the debilitating effect of relaxing damp climates, with all the attendant horrors, such as neuralgia, loss of appetite, and sleep, find that the dry, cold, tonic power of the air works the cure — appetite returns, digestion improves, "Nature's sweet restorer, balmy sleep," requires no sedatives to woo it, and the mind and body are left in a healthier condition.

In an ideal summer, like the last—1898—the air was lovely. Even in the hot months of August and September, when other places in the South and even in Scotland were sweltering in the heat, Harrogate was bearable and enjoyable. There is always what is known as a "kick" in the air here.

One sometimes hears that Harrogate is "such a windy place." Well, I cannot say that it is not. We certainly get a fair amount of wind, but it comes from a pure quarter, viz., from the Yorkshire hills, moors, and wolds. We know that ozone, an allotropic modification of oxygen, exists in the air in small

quantities, but in greater proportion in the country
than in confined cities, having the power of destroying
offensive odours, and is in itself a powerful bleacher
and intense oxidiser.

THE DRAINAGE.

The borough of Harrogate has a model system
of drainage and sewerage. The sewage matter and
surface water are conveyed to farms outside Harrogate—one called "The Jenny Plain Farm," 310 acres
in extent, lying along the Ripon Road on one side and
the Skipton Road on the other, and a second, beyond
Wetherby Lane—there on both farms to be deodor-
ised by the soil, and to fertilise the land. The sewers
have numerous ventilators, and consequently there
is an efficient escape of sewer gas, thus insuring a
healthy sanitary condition of the houses. Following
this, as a good result, is the almost entire absence of
typhoid fever, diphtheria, and ailments dependent on
foul air.

THE BOROUGH WATER SUPPLY.

Drainage, however effectual, must be accompanied
with a good domestic water supply. Fortunately
for all concerned, this Harrogate possesses.

There are two sources at present, viz., Beaver Dyke,
and Haverah Park. There are, in addition, two

service reservoirs, viz., Harlow Hill, and Irongate, Bridge Road. The sources being more or less removed from habitation, are collected from watersheds and moors, where little else is heard save the crowing of the cock grouse, the barking of the shepherd's dog, the bleating of the lamb, and the tinkling of its mother's bell. They are, therefore, perfectly pure and free from organic matter,—plentiful, and without hardness. It is not only for drinking and culinary purposes that the water supply is excellent. Its softness and purity make the morning ablutions a treat, and add whiteness and lustre to our linen. Those of us who revel in a "morning tub," know the luxury of floundering in its velvety liquidity, with no hardness to irritate the skin, but everything to soften, and make clear the complexion !

MORTALITY.

Harrogate has a *remarkably* low death-rate, and bears a most favourable comparison with that of other inland watering-places.

From all causes :—

> In 1894 it was 12·5 per 1000.
> „ 1895 „ 12·6 „
> „ 1896 „ 10·3 „
> „ 1897 „ 11·5 „

Average of 11·7 per 1000 per annum.

Deaths from infectious diseases (Zymotic):—

In 1894 it was ·48 per 1000.
„ 1895 „ ·05 „
„ 1896 „ ·05 „
„ 1897 „ ·02 „

Average of 0·15 per 1000 per annum. Comment on the above figures is needless.

As a Seat of Education.

The climate of Harrogate, on account of its bracing and dry character, to say nothing of the excellent drainage and water supply, already alluded to, is well adapted for children. Parents, therefore, may have every confidence in sending their boys and girls here. Several of the colleges for boys, though not what one would describe as " Public Schools," enjoy the reputation of claiming many, who, from being pupils, have gained high honours at Oxford and Cambridge, and who, in after life, have become men of eminence in the learned professions and in the public services.

There are also a large number of ladies' High class schools for boarders and day pupils, excellent in hygienic arrangements, and affording education of a firstrate order.

Socially Considered.

In addition to those who come in search of cure, it must not be forgotten that pleasant pastimes are

equally essential to those who "enjoy" life. In the
summer there is excellent fishing on the Wharfe
and Nidd; in the winter, "the mighty Nimrod" will
find the Bramham Moor and York and Ainsty fox-
hounds within rideable distance, two days a week, and
by train on others.

A new golf-course of 18 holes, with an excellent
club-house, is now to be found at Starbeck, and it
is only those who "do" a round who can speak
joyfully of the increased vigour which one feels
after it.

There is a good gentlemen's club, and clubs Con-
servative and Radical; a theatre, the Spa concert
rooms and gardens, where, in the season, the ear is
charmed and the mind soothed by the strains of an
excellent band: there are also lovely drives, in a most
interesting country.

GEOLOGICAL.

A few observations on the strata from whence the
various mineral waters of Harrogate are derived will
here not be out of place before describing the different
springs, and, without engaging the reader with any
lengthy scientific description, will doubtless prove of
interest to many.

GEOLOGY OF HARROGATE.

The mineral springs of Harrogate are, of course,
the peculiar product of the geological formation of

the district; and this patent fact desiderates at least an outline and sketch of the geological conditions which result in their production.

The carboniferous system, the uppermost system but one of the Palæozoic period, is the one which almost exclusively prevails in the Harrogate district. This system, from its base upwards, embraces the mountain-limestone, which is overlaid by the Yoredale series, the millstone grit series, and the coal-measures in succession. Hereabouts the coal-measures are wanting, and so the millstone grit overspreads the surface generally. *The principal exception to this arrangement is found in Low Harrogate, where the Yoredale series is uppermost, and which are the sources of the present day mineral springs.*

Here, as everywhere, the earth's crust is contorted and deflected into ridges and hollows, which are technically called "anti-clinals" and "synclinals," but in more familiar language we should call them ridges and hollows respectively; and the general direction in which these longitudinal folds run we designate the axial line. In the crumpling up of the rocks at the close of the Palæozoic period, and by the continuous effect of secular cooling of the earth's crust, an anti-clinal of the Yoredale series of beds has been forced upwards to the surface through the overlying covering of millstone grit at Low Harrogate, where this anti-clinal practically terminates at the surface, *and the mineral waters there are the outflow of the several strata of which this anti-clinal is composed;*

each of these strata *yielding water* **of** *a character* *peculiar to its own* *individual chemical nature, but* *quite distinct from* *that* **of** *its* *subjacent and super-* *incumbent neighbouring* *strata, which must,* **of** *course,* *be separated by impermeable* *intervening* *layers.* It is only an arrangement of this nature which could render possible the phenomenon of the various wells in the Bog Field, where we have a great number of wells of water of quite different composition standing closely side by side, and whose levels are not in the least affected by the pumping of water from any one or other of their neighbours. The fossilised life of these Yoredale rocks is *marine,* and from this and other facts we are justified in believing them to be *old sea-bottoms.* These conditions serve to invest a comparison of the analysis of sea-water and that of several of the Harrogate waters with special interest.

Tracing this anti-clinal backwards towards its origin in the Pennine Range, we proceed westwards, and at a distance of some 20 miles, in the Wharfe valley, we arrive at a point where the folds of the anti-clinal come to the surface; but here the folds are truncated, and their edges are arranged in such a manner as to permit the rains to penetrate the several strata, and so to gravitate, in this nature-formed conduit, down to Harrogate where the folds are similarly truncated, and thus yield their waters "for the healing of the nations," charged with the con-stituents of the beds which have conducted them to this highly favoured spot.

Analytical Report on Various Mineral Waters from Harrogate, forwarded by F. W. Smith, Esq., M.D., during October, November, and December, 1898.

THE gases were collected and the constituents liable to oxidation were estimated at Harrogate within an hour or two of the sample of water being collected, with the exception of the gases in the first sample of Kissingen water.

The results are given in tabular form in grains per gallon. As one grain per gallon (1 in 70,000) is equivalent to 0·0014 per cent., I have not thought it advisable, in framing this report, to state the results beyond one-tenth of a grain (0·00014 per cent.), except in the case of one or two rare but important constituents.

CHARLES F. TOWNSEND,

Fellow of the Chemical Society, Member of the Society of Chemical Industry, Late Assistant Examiner Royal College of Physicians, London.

THE LABORATORY, CHARING CROSS HOSPITAL, LONDON, 2nd *January* 1899.

	Old Sulphur Well, Royal Pump Room.	Strong Sulphur, Montpellier.	Magnesia Well.	Well No. 36.
Sodium, . . .	340·6	252·6	61·2	95·1
Potassium, . . .	13·4	2·1	3·6	13·8
Lithium, . . .	·14	·12	...	·08
Ammonium, . . .	·23	·17	·06	·12
Barium, . . .	3·6	2·3	·8	1·2
Strontium, . . .	Trace	·9	Trace	Trace
Calcium, . . .	28·9	21·6	4·9	7·4
Magnesium, . . .	11·7	9·8	3·0	2·9
Iron, . . .	...	...	...	
Manganese, . . .	...	...	...	...
Chlorine, . . .	602·9	449·8	99·4	159·1
Bromine, . . .	1·7	1·3	Trace	Trace
Iodine, . . .	Trace	...	...	...
Sulphur in Sulphides, .	·34	1·6	·4	·6
Sulphur trioxide, . .	...	·24	...	·3
Carbon dioxide, . .	14·0	5·9	9·8	14·0
Oxygen of bases, . .	5·1	2·5	3·5	5·2
Silica, . . .	2·8	·7	1·0	·7
Totals, .	1025·4	751·5	187·6	300·5

Gases in Cubic Inches per gallon.

Sulphuretted Hydrogen,	16·0	11·5	...	5·6
Carbon dioxide, . .	30·5	31·9	21·0	30·5
Nitrogen, . . .	6·9	4·4	7·2	...
Oxygen, . . .	...	...	...	...
Marsh gas, . . .	...	2·8	...	...
Totals, . .	53·4	50·6	28·2	36·1

The quantities are expressed in

	Chloride of Iron Spa.	Kissingen Spa, 1st Analysis, Collected 6 P.M.	Kissingen Spa, 2nd Analysis, Collected 6 A.M
Sodium, ...	151·9	91·6	184·7
Potassium, ...	1·9	·7	1·1
Lithium, ...	·03	Trace	·3
Ammonium, ...	·12	·1	·4
Barium, ...	2·0	1·1	2·0
Strontium, ...	·4	Trace	·2
Calcium, ...	19·2	14·2	25·3
Magnesium, ...	5·7	4·0	10·4
Iron, ...	11·2	·11	2·3
Manganese, ...	Trace	...	...
Chlorine, ...	293·2	161·5	345·4
Bromine, ...	·5	Trace	1·1
Iodine, ...	...	...	...
Sulphur in Sulphides, . . .	...	...	...
Sulphur trioxide, ...	Trace	·5	·8
Carbon dioxide, ...	5·9	11·3	12·1
Oxygen of bases, ...	2·1	4·2	4·4
Silica, ...	1·1	2·4	3·5
Totals, . .	495·2	291·7	594·0

Gases in Cubic Inches per gallon.

	Chloride of Iron Spa.	Kissingen Spa, 1st Analysis	Kissingen Spa, 2nd Analysis
Sulphuretted Hydrogen, . . .	...	...	...
Carbon dioxide, ...	9·4	23·0	25·8
Nitrogen, ...	6·1	4·1	6·6
Oxygen, ...	Trace	·5	·6
Marsh gas, ...	...	...	...
Totals, . .	15·5	27·6	33·0

CHARLES F. TOWNSEND, F.C.S., &c.

grains and cubic inches per gallon.

TABLE OF ANALYSES OF SULPHUR WATERS, comparing other British and

The chemical terms are those employed
The quantities are expressed in grains

CONSTITUENTS.	Old Sulphur Well, Royal Pump Room (Townsend).	Strong Sulphur, Montpellier (Townsend).	Magnesia (Townsend).	Well No. 26 (Townsend).	New or Mild Sulphur, Royal Pump Room (W. A. Miller).	Mild Sulphur, Montpellier (Attfield).
Calcium Sulphuret,	..	..	..	..	..	..
Magnesium Sulphuret,	..	..	..	..	..	..
Sodium Sulphydrate,	·6	2·8	·8	1·1	6·89	..
Sodium Sulphide,	..	..	..	..	..	8·777
Barium Chloride,	5·5	3·5	1·2	..	Trace	..
Strontium Chloride,	Trace	1·7	..	Trace	..	·619
Calcium Chloride,	44·92	45·1	..	..	16·70	31·296
Magnesium Chloride,	46·31	38·8	2·3	..	2·39	27·589
Potassium Chloride,	23·0	4·0	6·9	26·4	11·34	5·691
Lithium Chloride,	·8	·8	..	·5	Trace	..
Ammonium Chloride,	·7	·2	·2	·4	..	·656
Sodium Chloride,	865·3	638·1	154·9	241·6	82·95	388·800
Magnesium Bromide,	..	..	..	..	..	..
Sodium Bromide,	2·2	1·6	..	Trace	..	..
Magnesium Iodide,	..	..	..	..	..	..
Sodium Iodide,	Trace	..	..	..	..	..
Calcium Carbonate,	31·83	13·4	12·3	11·1	..	16·711
Barium Carbonate,	..	..	..	5·3	..	..
Magnesium Carbonate,	..	..	8·3	7·2	..	..
Potassium Carbonate,	..	..	..	..	..	..
Sodium Carbonate,	..	..	..	..	..	..
Magnesium Sulphate,	..	..	..	..	..	..
Aluminium Sulphate,	..	..	..	..	..	..
Strontium Sulphate,	..	..	..	..	..	·913
Potassium Sulphate,	..	..	..	..	..	..
Calcium Sulphate,	..	..	..	..	..	..
Sodium Sulphate,	..	·4	2·8	·6	..	..
Sodium Nitrate,	..	..	..	..	..	·370
Calcium Phosphate,	..	..	..	..	..	..
Phosphates of Alumina and Lime	..	..	..	..	..	..
Calcium Fluoride,	..	..	..	..	..	..
Silica,	2·8	·7	1·1	·7	2·40	3·836
Iron Peroxide (sesquioxide),	..	..	..	..	..	..
Alumina,	..	..	..	..	..	..
Silicic Acid,	..	..	..	..	..	..
Sulphur in suspension,	..	..	..	..	..	..
Totals,	1025·4	751·5	187·6	300·4	654·87	485·258

Gases in

	Old Sulphur Well	Strong Sulphur	Magnesia	Well No. 26	New or Mild Sulphur	Mild Sulphur
Sulphuretted Hydrogen,	16·0	11·5	..	5·6	4·18	..
Carbon dioxide,	30·5	31·9	21·0	3 0·5	13·22	54·00
Carburetted Hydrogen,	..	2·8	..	..	..	·8
Nitrogen,	6·9	4·4	7·2	..	2·01	3·26
Oxygen,	..	..	..	..	..	..
Totals,	53·4	50·6	28·2	36·1	19·41	58·09

by the various analysts.
and cubic inches per gallon

Starbeck (Fairley).	Llandrindod Wells (Wanklyn).	Llanwyrtyd.	Moffat.	Strathpeffer.	Schinznach (Granteau).	Eilsen (Dunesnil).	Aix-la-Chapelle.	Weilbach.	Aix-les-Bains.	Baden (near Vienna).
..	..	..	..	..	·560	..	..	..	..	..
..	..	..	..	..	..	..	..	..	..	3·68
1·36	..	..	..	..	..	..	·952	..	..	..
Trace	..	..	..	..	..	..	..	..	..	..
..	47·2	13	..	..	..	..	..	..	..	..
..	5·6	..	..	..	..	1·47	..	..	1·20	..
..	..	..	..	..	6·020	..	..	2·13	..	..
Trace	..	..	..	..	..	..	..	..	..	..
Trace	..	..	..	..	..	..	..	..	..	..
116·44	235·9	61	4·5	..	40·950	..	184·24	20·83	·55	20·41
Trace	..	..	..	..	..	..	..	..	..	..
..	..	..	..	..	..	..	·252	..	..	..
Trace	..	..	..	..	..	..	·035	..	..	..
10·01	7·3	..	..	..	17·50	21·26	..	29·09	10·39	..
3·51	..	..	..	..	8·4	1·47	..	27·58	1·81	..
·65	..	..	..	..	..	..	..	..	..	..
14·47	..	..	..	..	..	..	45·325	31·32	..	7·49
..	..	..	..	32	..	45·66	..	..	2·46	..
..	..	..	..	..	..	..	..	..	3·83	..
..	..	..	..	..	..	..	..	2·98	..	5·83
1·88	1·0	..	..	51	73·670	139·30	..	..	1·12	11·43
..	..	..	..	..	..	53·07	19·85	..	6·72	24·1
..	..	..	..	..	..	11·79	..	..	..	..
..	..	..	..	..	..	..	..	..	·174	..
3·27	..	..	..	..	·770	·060	..	1·11	..	2·85
..	..	..	..	..	·35	·061	..	..	..	..
..	..	..	..	..	·700	..	..	..	..	..
..	..	..	..	4	..	..	..	..	·35	..
151·59	297·0				148·92	273·48			30·10	

cubic inches.

..	"more than 1 in 1000"	10	8	11·26	10·55	14·33	..	1·6	7·3	·8
..	..	..	..	10·55	..	13·06	..	31·2	3·60	14·3
..	..	..	..	..	..	·75	..	..	..	..
..	..	..	..	..	..	2·88	..	..	6·99	..
..	..	..	..	..	..	·71	..	..	..	..
..	..	..	..	..	..	31·73	..	32·8	17·89	15·1

C. F. Townsend, F.C.S., together with two typical

The chemical terms are those employed
The quantities are expressed in grains

	Kissingen (Townsend), 1st Analysis, Collected at 6.0 p.m.	Kissingen (Townsend), 2nd Analysis, Collected at 6.0 a.m.	Chloride of Iron Water (Townsend).
Ferrous Chloride,	·3	5·2	8·4
Ferrous Carbonate,	..	..	15·5
Ferrous Sulphate,	..	..	..
Ferric Sulphate,	..	..	..
Iron Peroxide,	..	..	..
Bicarbonate of Protoxide of Iron,	..	..	..
Aluminium Sulphate,			
Calcium Sulphate,			
Magnesium Sulphate,			
Potassium Sulphate,	..		
Ammonium Sulphate,			
Barium Sulphate,	..		
Sodium Sulphate,			
Strontium Sulphate,			
Potassium Chloride,	1·3	2·1	3·6
Sodium Chloride,	253·1	468·5	386·5
Ammonium Chloride,	·2	·8	·3
Barium Chloride,	1·4	2·8	3·0
Strontium Chloride,		·3	·7
Calcium Chloride,	10·8	39·7	53·2
Manganese Chloride,			
Magnesium Chloride,	16·0	41·1	22·5
Lithium Chloride,	..	..	·2
Lithium, Bromide, Iodide and Fluoride,	..	..	..
Magnesium Iodide,	..	..	
Sodium Iodide,	..	..	
Magnesium Bromide,			..
Sodium Bromide,	Trace	1·4	·7
Barium Carbonate,			
Calcium Bicarbonate and Carbonate,	25·8	27·5	..
Magnesium Bicarbonate and Carbonate,	..	..	
Potassium Carbonate,			
Sodium Bi- and Carbonate,	..		
Manganese, Bicarbonate of Protoxide of,	..	..	
Cobalt, Nickel, Bicarbonate of Protoxide of,	..	..	..
Lithium Bicarbonate,	..		
Strontium Bicarbonate,			
Potassium Nitrate,	..	..	
Sodium Nitrate,		..	
Calcium Phosphate,			
Aluminium Phosphate,	..	..	
Silicic Acid,			
Silica,	2·4	3·5	1·1
Organic Matter,	..	..	
	291·7	592·9	495·7

Gases in

Carbon dioxide,	23·0	25·8	9·4
Carburetted Hydrogen,			
Oxygen,	·5	·6	Trace
Nitrogen,	4·1	6·6	6·1
Sulphuretted Hydrogen,	..	..	
Ammonia,		..	
	27·6	33·0	15·5

by the various analysts.
and cubic inches per gallon.

Alexandra Chalybeate (Davis).	Carbonate of Iron Spa (Musprat).	Pure Chalybeate, Royal Pump Room (Davis).	Tewit Well (Hofmann).	John Well or Old Spa.	Alum Well (Davis).	Homburg, Stahlbrunnen.	Kissingen Rakoczy (Liebig).
5·8	0·042	1·364	1·358	1·271	..	..	2·18
..	..	..	..	..	69·83	..	..
..	..	..	..	..	78·76	..	..
..	..	..	..	..	..	6·885	..
..	..	..	..	..	89·47	..	..
9·007	7·625	·749	·697	·307	56·91	·261	25·93
..	..	..	..	..	57·33	..	41·01
..	..	..	..	..	3·14	..	..
..	..	..	..	..	2·19	..	..
..	..	..	..	..	..	·029	..
..	..	..	..	..	..	·742	..
1·130	·150	..	1·323	..	..	17·38	20·05
176·370	11·650	1·625	·280	1·543	33·95	410·42	407·51
Trace	..	Trace	Trace	Trace	..	·91	..
.	..	..	..	..	..	..	..
..	2·311	..	..	..	..	34·84	..
Trace	..	Trace	Trace	..	..	..	..
4·736	13·148	..	..	..	..	22·08	21·23
..	..	..	..	..	..	·84	1·36
Traces	..	Traces	Traces	..	..	..	..
..	..	..	..	..	..	·001	..
..	..	..	..	..	..	..	Trace
..	..	..	..	..	..	·047	..
..	..	..	..	..	..	..	·54
13·762	·341	1·532	1·435	2·364	..	72·80	74·19
5·785	..	1·952	2·667	3·039	..	6·54	..
..	..	·262	1·057	·991	..	..	..
..	..	1·103	..	1·338	..	..	..
..	..	..	..	..	..	·392	..
..	..	..	..	..	..	·002	..
..	..	..	..	..	..	..	..
..	..	..	..	..	..	·131	..
..	..	..	..	..	..	..	·63
...	..	..	..	..	..	·071	·36
..	..	..	..	..	..	12·032	..
·675	·204	·502	1·041	Trace	3·27	..	·82
1·450	..	·750	·663	Trace	..	..	..
218·804	41·471	9·859	10·521	10·753	394·41	..	585·8

cubic inches.

17·04	..	13·74	11·85	14·95	..	656·5	380·7
..	..	..	..	·15	..	..	..
·31	..	·82	·4	·67	..	..	..
8·98	..	8·00	5·53	6·55	..	..	..
..	..	..	..	..	..	·12	..
..	..	..	..	..	..	..	·6
26·33	..	22·56	17·78	22·12	..	..	381·3

THE MINERAL WATERS.

That there should be something like eighty mineral springs within a compass of two miles in and around Harrogate, most of them more or less different, is a thing "which no fellow can understand." But such is the case. Sir W. Slingsby I have alluded to as the discoverer of the Tewitt Well, to Dr Deans and to Dr Stanhope, pioneers in discovery after him. A century after Dr Stanhope wrote his work, Dr Short, in 1734, added another. But it does not appear that even then the number of wells was more than a dozen. However, as time went on, fresh discoveries were made, and there is no knowing what may be in store for future explorers; let us all work on. The Bog Fields account for the greatest number of springs; after them, what are known as Montpellier Gardens (now the Royal Baths), Royal Pump Room and surroundings, and Spa Pump Rooms.

Outside Harrogate are the sulphur wells of Beckwith, Bilton, Starbeck, and Harlow Car.

Having given the geological features and the strata, in which the Harrogate mineral waters are found, I will now proceed to consider the different groups of waters themselves.

And before going further, let me ask, are they the same in consistency and salts they were twenty years ago? From the fresh analysis some are, others are not quite alike. For sufficient reasons, last season, I was sceptical about the Kissingen water, drawn at

different times of the day. I now find that this
water varies. If drawn in the morning it is about
the old strength; if later in the day, it is less charged
with both the aperient elements, less also with the
restorative hæmoglobin ingredient—iron. Therefore,
I advise patients "to go for" the morning "tap." If
taken in the afternoon the dose must be increased.
The chloride of iron water, too, is somewhat erratic;
at one time the chloride of iron is in excess, at others
the carbonate. But this variation is not very
material, so long as there is no diminution of the
salts, in either form.

From the different analyses, there appears to be
no diminution of the total salts.

On looking at the sulphur group, the new analysis
has made certain revelations. The reader, however, will have all that explained later on at
page 39, and he will be able to see what the
changes are.

Up to the present all has been explanatory, and
the last chapter "dry bones."

But now I feel like the spirited hunter from
County Meath, who, with all his Irish blood coursing in
his veins, when he hears the "tally ho," and, sighting
the red coats streaming across the Emerald pasture,
"shifts place and paws" to follow in the flight; or
like the salmon fisher who, for days together, has
threshed the mountain stream without "a rise,"
when suddenly he receives the unexplained thrill,
and lion-like the silvery beauty comes with a dart

giving the electric quiver "all along the line," now
rushing up stream, now down, here "sulking," there
lolloping in the air, till at last, weary and worn out
with his endeavour, he is "grassed" by the stalwart
Angus, to the tune of the latest "lament," and "a
wee drappie" of peaty Glenlivet.

But some will say, What has this got to do with the
case? When we consider the facts, and what is in
store, who is there who will not ask: Is sulphur not a
strong enough subject to become excited over? But
this in passing.

Before grouping the different waters together I
will explain what the sulphur water of Harrogate is,
and from whence it derives its odoriferousness.

In the minds of many, Harrogate waters are only
associated with "a bad smell"—"rotten eggs" and
odours appertaining thereto.

It is the presence of sulphuretted hydrogen gases,
which, together with sulphide of sodium and sulphy-
drates, gives the *sulphur odour* and the *sulphur metal*.
In addition to their value medicinally, they may be
said to be the "gold mine" of Harrogate. What
would not Leamington give to have a sulphide in its
mineral waters? They would "smell as sweetly as
those of Harrogate"—perhaps they will some day,
should a volcanic outbreak shake the land of Shake-
speare, and alter the strata.

To make a mineral water popular with the public,
it must "do" something very much out of the way.
It must either "smell nasty," or come hot from the

ground. This may seem strange, but it is not the view taken by the physician. And when on this subject of "odours sweet," let me remark that it is usual with authors in writing books on Harrogate and Strathpeffer to vie with one another in trying to prove which is stronger in sulphuretted hydrogen.

In the new analysis by Mr Townsend, some very striking facts come out.

Take, first, the gases in the old sulphur well. They are now proved to be 16·0 cubic inches of sulphuretted hydrogen per gallon, equivalent to 6·1 grains of sulphur in the gallon, as against 10·16 cubic inches of sulphuretted hydrogen in the analysis of Thorpe, 1875.

This settles the question, which has been a long and vexed one, viz., that in the old Harrogate sulphur well we have **the strongest sulphur well in Europe**.

Secondly, in the strong Montpellier sulphur water, Mr Townsend found 11·5 cubic inches of sulphuretted hydrogen gas, equivalent to 5·7 grains per gallon, as against nil in *Professor Attfield's* analysis of the same water, 1879.

That these facts are correct I have not the slightest doubt.

The gases were collected, and the constituents liable to oxidation were estimated, at Mr Eynon's laboratory, by Mr Townsend within an hour or two of the time they were drawn from the respective wells.

Fortunately the new analyses show up the Harrogate waters in a stronger and better light, in almost every particular.

There is one well, the Kissingen, which calls for remark. As already observed, the strength of the salts in this water varies according to the time of day at which it is drawn, and it is possible this may have always been the case.

Mr Townsend examined two samples; the first was taken from the well at 6 p.m., the second at 6 a.m. The difference the reader will see is very remarkable, and it is to be hoped will be able to be remedied. I am sorry to have to record this fact, *but as I had the various wells analysed, at a great expense to get at the truth,* I am bound to record it " without fear, favour, or affection."

Such, then, is the most recent analysis of six mineral waters from as many wells, mostly in use in Harrogate, for medicinal drinking purposes.

I will now proceed to arrange the natural waters of Harrogate in their various groups.

I. SULPHUR.

These are two—

A. Sulphur Alkaline.

Beckwith Water,
Harlow Car,
Starbeck,
Bilton,
} These lie outside Harrogate.

B. Sulphur Saline.

Old Sulphur Water (Royal Pump Room).
Strong Sulphur Montpellier (Royal Baths).
Mild Sulphur Montpellier (Royal Baths).
New Sulphur No. 36 Water (Valley Bog Fields).
Magnesia Water (Valley Bog Fields and Royal Pump).

II. CHALYBEATE WATERS.

A. Non-Saline Chalybeates.

Tewitt Waters (Stray).
St John's Waters (Stray).
Pure Harrogate Chalybeate (Royal Pump Rooms).
Carbonate of Iron Water (Spa Pump Rooms).

B. Saline Chalybeates.

Kissingen (Royal Baths).
Chloride of Iron (Spa Pump Room).
Alexandra Waters (Royal Pump Room).

C. Chalybeate Sulphated (Ferric) Water.

Alum Well (Bog Fields).

I. SULPHUR WATERS.

In the sulphur alkaline groups, viz., Harlow Car, Starbeck, Bilton, and Beckwith, carbonates of sodium and potassium are found as alkalies. This is accounted for by the absence of chlorides of magnesium and calcium.

In the sulphur saline waters, chlorides of sodium, potassium, magnesium, calcium, barium, strontium, and lithium, make up the principal saline salts.

It will be seen on examining the tables of comparison of the waters of Harrogate with those of Continental Spas, and also with others of this country, that all the salts above mentioned compare favourably.

BARIUM.

The importance attached to this metal, in recent years, as a therapeutic agent is such, that I cannot help giving it a prominent place at once. Barium was found, according to Mr Townsend's analysis, in all the six wells, which he analysed in a greater or less degree.

In 1866, Mr Haydon Davis, F.C.S., found barium and strontium in five wells. He has been kind enough to allow me the benefit of hearing in conversation all about his interesting calculations. This discovery and these analyses have been verified by all subsequent analysts.

The presence of barium chloride in six of the Harrogate wells, analysed recently by Mr Townsend, is a fact that cannot be over-estimated. It is a salt which is present in considerable quantities in colliery waters, and is generally associated with chlorides of calcium, sodium, and lithium. It is present in a spring near Shotley Bridge, a village in Durham, in quantity of 4 grains per gallon, and for many years has been used in cases of rheumatism and skin diseases. It is also found in Llangammarch waters in Mid Wales to the extent of 6·749 grains to the gallon.

Its value in therapeutics is that its properties are similar to those of digitalis. It produces muscular contraction, and contracts the blood-vessels; it also stimulates the cardiac muscles and the capillaries.

In 1888 Dr A. Bary, of Dorpat, made a series of valuable investigations with this drug, and he found that its chief action was on the heart, and was very similar to that of digitalis. The chloride, he says, "increases the action of the heart muscle, in small doses, and in larger ones sets up peristaltic movements, arresting the heart finally in systole."

Dr Hare, in *Philadelphia Medical News*, 1889, p. 183, found also that barium chloride slows the heart, steadies its rhythm, and increases the volume of blood thrown out of the ventricles; similar results, also, have been found by Lauder Brunton, Ringer, and Robart.

It is said by some that chloride of calcium assists barium chloride in its action. Both salts act in a similar way on the vascular system—that is, they strengthen the heart's action, and contract the peripheral vessels. Lauder Brunton, however, found that they were antagonistic to each other, for whilst the action of barium is rapid, that of calcium is slow, and when the latter was given after the former, the barium pressure curve was reduced to normal by the calcium before the calcium gradual curve was produced.

These experiments, of course, were made with salts, not as combined in mineral waters, and whether the antagonism exists in the Harrogate waters has not, so far as I am aware, been as yet determined.

In any case there is no doubt that barium chloride is a most valuable constituent in Harrogate waters.

ABSENCE OF SULPHATES.

It is remarkable, but nevertheless true, that there is an almost entire absence of sulphates of sodium, magnesium, and calcium, in the Harrogate sulphur group. Perhaps a little of the former two, sodium and magnesium sulphates, would have made them more valuable.

II. CHALYBEATE WATERS.

(*a*) Ferrous Carbonate.
(*b*) Ferrous Chloride.
(*c*) Ferrous and Ferric Sulphates.

I have already indicated where these waters are found, and have given their names.

Of the seven wells, the Kissingen and chloride of iron waters are mostly drunk medicinally, because of the peculiar combination of their salts. This will all be gone into later on.

In each well, whether sulphurous or ferruginous, will be found a simple prescription. It is well known in medicine *that the action of mineral waters is different in character and degree from the pharmaceutical solutions of their ingredients.* This is especially borne out by Sir Henry Thompson, in his remarks upon the action of salines in gravel.

It has been a fashion to imagine that the stronger a mineral water is in aperient properties, the more likely it will be to produce favourable results. This is a fallacy. For a prolonged course, it is better that it should not be so. However, in Harrogate, such is the variety of springs themselves, and strength of composition, that almost every peculiarity of constitution can be accommodated.

PHYSIOLOGICALLY AND THERAPEUTICALLY CONSIDERED —EACH BASE SEPARATELY DEALT WITH.

Before arriving at the general conclusion, as to the action of the different groups as a whole, I purpose to bring before my readers, *first*, the bases of the salts, and their action; *second*, the salts themselves as they are separately combined in the waters; *third*, what the real action of the waters is, from the time they enter the body as a whole, till they leave it by the bowels, kidneys, etc.; their consequent influence on the tissues and the blood; and finally, the ailments indicating their administration.

Before going further I may remark that I shall only touch lightly on the various bases and salts and their actions. I will make my remarks brief and succinct, for this is not an elaborate work on Materia Medica and Therapeutics.

I. SODIUM.

Uses.—Internally sodium salts are more slowly absorbed into the system than those of potassium, and they are therefore more powerful in the alimentary canal, in Harrogate water. Sodium reaches the stomach in the form of chloride mostly, and assists the digestion of albumen ; and it must not be forgotten that bicarbonate of sodium, when received by the same organ, is in part converted into chloride.

Action on the Blood.

The salts of sodium are slowly absorbed into the blood, and are slowly excreted from it, remaining in it chiefly as the bicarbonate and phosphate. They are taken with our daily food, and are the chief sources of the natural alkalinity of the *liquor sanguinis.* This alkalinity may be increased by their being given in the form of medicines, or of mineral waters containing chloride or bicarbonate of sodium, as well as other sodium salts.

The fact that sodium salts are alkalisers of the blood is utilised in gout, rheumatism, gravel, and various disorders and diseases of the liver.

Remote Actions.

Sodium salts are excreted by all mucous surfaces, kidneys, liver, bronchi, and probably by the skin. During their passage the functional activity of these organs is increased.

The different salts of sodium have different actions —some affecting one organ, some another, but all are ultimately alkalisers.

When, therefore, we sum up the "all round" action of the sodium salts, we say—they act on the alimentary canal, blood, tissues, and on the organs and passages of the body by which they are excreted.

They are, therefore, indicated in all those whose unfortunate bodies are known as "gouty," "rheumatic," "acid," or where there is chronic functional derangement of the liver.

Chloride of sodium possesses the ordinary action of sodium, and it is greatly used in the diathesis above mentioned in the waters of Homburg, Wiesbaden, Kissingen, and Baden-Baden.

II. MAGNESIUM.

Internally, magnesium has the power of decomposing the contents of the stomach and intestines under different circumstances. By this process it neutralises ordinary, or too abundant acidity in these organs, and is itself converted into the chloride, lactate, and bicarbonate, thus removing irritant acids, and forming salts of magnesium, which have a stimulant or purgative action on the intestines.

The chloride of magnesium as present in the Harrogate waters is slowly absorbed, and produces local effects as a *saline purgative*.

Action on the Blood.

Magnesium enters the circulation as the chloride, increases the alkalinity of the plasma, of which it is a normal constituent, and helps to hold in solution any acid which may be in excess. It will, therefore, be found useful in gout, lithiasis, and chronic rheumatism (Mitchell Bruce).

Remote Action.

If magnesium does not act as a purgative, it is excreted chiefly by the kidneys, rendering the urine more abundant and less acid. Its diuretic and alkalising effects contribute to the value of magnesium waters in gout and gravel (Mitchell Bruce).

III. CALCIUM.

Internally, calcium is an antacid. It increases the alkalinity of the blood, and in the form of chloride, as it exists in some of the Harrogate waters, is strongly recommended in scrofulous diseases of glands, in tuberculosis, and anæmia.

Calcium salts have also a marked diuretic power, hence the value of waters containing them in such affections as gravel, gout, and rheumatism.

AMMONIUM.

Uses.—Taken internally, Ammonia is a most powerful general stimulant. In cases of cardiac failure, it increases the heart's action, and resuscitates feeble respiratory movements.

Action on the Blood.

Ammonia is an alkaliser of the blood plasma, and prevents it becoming coagulated under certain circumstances. The phosphate keeps uric acid in solution, and is therefore advantageous in gouty affections.

Remote Actions.

The chloride acts as a diuretic in dropsy; is given in certain liver diseases, and is an excellent expectorant in cases of bronchitis.

POTASSIUM.

Potassium is an alkaliser of the blood, and its qualities physiologically and therapeutically will be dealt with later on ; so too lithium.

Physiologically and Therapeutically Considered —Each Salt separately dealt with.

Chloride of Sodium.

This salt is found in greater quantity in the sulphur group than any other. It is found in the human body as part and parcel of it : and if through disease its presence be reduced, it is made up, if taken as a medicine, in this group.

Dilute solutions of chloride of sodium have the power of dissolving albumins and globulins, while strong solutions precipitate globulins and withdraw water from the tissues.

Chloride of sodium therefore acts on the gastric glands in proportion to the strength of its administration. If the dose be weak it acts as a mild stimulant ; if strong, the action is irritant and the result is sickness or purgation.

During its stay in the body, the salt does not seem to alter the composition of the tissues ; but what change takes place seems due to its action on the solubility of albuminous substances, on the processes of osmosis between the intercellular fluid and the blood, and the circulation of lymph in the tissues.

Therefore chloride of sodium increases tissue change by an increase of the amount of urea excreted. It is not always present in the body in the same proportion. According to the quantity taken daily, so is the balance kept up, more or less, although it may take two or three days to become apparent.

Chloride of sodium stimulates the construction of tissues generally and retards retrograde metamorphosis.

In addition to all these properties, its value is enhanced by belonging to the sodium group.

CHLORIDE OF CALCIUM.

In my introductory remarks I said that in the early history of the Harrogate waters, people drank them "chiefly as an antidote to scrofulous affections." Although the particular healing salt was probably not differentiated, nevertheless I have little doubt that chloride of calcium, in combination with others, was the one in which the virtue existed.

Chloride of calcium is found in considerable quantity in several of the sulphur, as well as in many of the iron, group of waters.

In the present day it is largely and scientifically prescribed in strumous affections, and in them its use is of considerable value.

Dr Robert Bell (*The Lancet*) finds that it possesses wonderful power in controlling, if not actually curing, many forms of tubercular disease; and in the wasting of childhood he has found it a therapeutic agent of considerable value. Dr J. G. S. Coghill (*The Practitioner*) regards chloride of calcium as possessing quite the character of a specific in strumous disease—more potent and more manageable than preparations of iodine. All this I have found to be true: chloride of calcium, therefore, assists the sodium chloride in preventing retrograde metamorphosis of tissues ; it stimulates the reconstruction of them generally, and has a peculiarly remedial action upon glandular substances, as is seen in struma, tabes mesenterica, bronchocele, certain stages of tuberculosis, and in enlarged tonsils. *It also strengthens the heart's action, and contracts the peripheral vessels.* Chloride of calcium has thus at the present time found favour in the treatment of tubercular disease after being discarded for sixty years ! !

CHLORIDE OF MAGNESIUM.

This salt is a laxative. It acts freely on the bowels, and its use is not likely to be followed by constipation. It does not seem to continuously irritate the gastric and intestinal glands, and hence its administration does not " fidget " the patient with repeated desires for action.

CARBONATES OF MAGNESIUM AND CALCIUM.

These are met with in considerable quantities in some of the wells—more especially in the mild Montpellier and in the magnesia wells in the bog fields. They more or less act as alkalisers and remove acidity—the one more actively, as in the case of the magnesium carbonate, and the other in a less degree. They both, however, operate in the right direction, and are most valuable in gout, rheumatism, and gravel, and allied conditions of ill health.

SODIUM BICARBONATE.

All that has been said under the heading sodium applies to this salt. It is found in the waters of Starbeck, Beckwith, Harlow Car, and of Bilton.

In addition to its value internally, it is most useful in the waters of these wells for bathing purposes, in such ailments as gouty eczema, psoriasis, etc. The waters of these wells just enumerated contain little or no chlorides, but everything that is soothing and nothing irritating in them.

Potassium, lithium, sodium iodide, and sodium bromide are found in more or less quantities in some of the wells and enhance their value. When we consider how Continental physicians extol waters of foreign spas containing even a trace of these metals, I think I am justified in calling attention to their presence in those of Harrogate. All are alkalisers and more or less antidotes to gout.

THE SULPHURETTED HYDROGEN GASES, with the addition of sodium sulphide and sulphydrate, as found in the strong sulphur waters of Harrogate, assist as purgatives by increasing the peristaltic actions of the intestines, by stimulating the intestinal glands. They thus act on the liver through the portal circulation, especially when the blood is gouty or apoplectic with hæmoglobin in excess, and assist that organ in its destruction.

Action on the Blood.

As they enter the blood, they do not seem to produce any specific effect upon it. Whatever action takes place in the tissues, through the blood, is on the central nervous system, and, in this way, they possess alterative powers.

The sulphides pass through the kidneys as sulphates. They also pass through the skin as sulphides. Outwardly applied as baths they are specifics in certain forms of skin disease.

Tracing, therefore, their uses and actions, in such ailments as chronic gout, rheumatism, diseases of the skin, and engorgement of the chylopoietic viscera, they relieve the portal circulation by purgation, and stimulate the kidneys and skin; they are also exhaled in the breath, and are useful in chronic affections of the respiratory tract.

TEMPERATURE OF THE WATERS.

All the sulphur waters of Harrogate are cold. The principal class of these waters, I mean the sulphur group, is Harrogate (England), Strathpeffer and Moffat (Scotland), Llandrindod (Wales), Gurnigel (Switzerland), Heustrick (Switzerland), Nemidorf (Prussia), and Weilback (Germany).

The chalybeate waters of Harrogate are also cold.

THE COMBINED ACTION OF THE SALTS
IN THE SULPHUR GROUP.

The question arises, What is the combined action of this group in the Harrogate waters? I have tried to indicate the action of the salts separately. I will now gather up the crumbs, and put them together.

The word sulphur is derived from *sal*, salt, and πυρ, fire—an element of bivalent combining power —brimstone or burntstone (Saxon), and its uses have a very ancient history. I believe they are spoken of in the Bible. At any rate I have vivid recollections of a wholesome dread, as a youth, of what I was likely to expect in the next world, " if I continued to be a naughty boy." Fire and brimstone were in store for me, probably for purifying purposes.

In later years who is there whose blood has not curdled at the heart-thrilling tale of Dickens, where brimstone and treacle combined with starvation were daily meted out to the skinny urchins, in Squeer's School, at " Do-the-Boys'-Hall " for what wise purpose was best known to that gentleman himself.

Action—internally, purgative and diuretic ; they are liver stimulants, alkalisers, and general blood purifiers.

The water from the old sulphur or from the strong sulphur Montpellier well, taken warm early in the morning, in quantity from twelve to twenty-four ounces, followed by a gentle walk, hot tea and breakfast, produces a smart, liquid motion, with a feeling of relief of weight from the abdominal cavity generally.

In many cases, the "first attempt" of drinking is followed by an inclination to vomit, but this is generally got over in a day or two. It is really, in most cases, the "smell" that brings this about, not actually the taste, and it may seem strange, but after a time patients often get to like the water. When the stomach successfully retains the draught, not infrequently a slight sense of oppression is experienced, which quickly passes off; at times, also, a sense of determination of blood all over the bowels with distension. This feeling, however, generally gives place to a subdued sense of "settling down," and very little else is felt till the well-known twinge and general quiver announce that an effect is to be expected, and the patient is seldom disappointed. Copious, free, liquid, easy purgation is almost invariably the result, and a general sense of lightness is felt afterwards.

The exact manner in which purgation by saline cathartics is brought about has been a puzzle to physicians, but what with the researches of Rutherford, Matthew Hay, Lauder Brunton, and others, our knowledge on the subject has become more exact.

It is not intended in this little work to enter fully into the subject, and the reader is therefore referred to Dr Matthew Hay's masterly work on the subject.

On examining the faeces after a strong dose of the water from the old sulphur well, it will be found from their bilious character that the liver has been freely acted upon. I have already given the explanation of this, when I stated that the liver is stimulated by the sulphuretted hydrogen and sodium salts. There is an old saying and a very true one, " sodium for the liver, potassium for the blood." It is generally found that patients bear this free purgation in Harrogate, in a wonderful way. I cannot help thinking that this is due to the stimulating effect of the chlorides of calcium and ammonium in the old sulphur well waters, with the addition of the barium. These salts, as I have explained, " support " the heart, and patients, therefore, do not complain of the exhaustion which follows the operation by *strong sulphated mineral* waters. Of course, it must not be forgotten that the pure and bracing air of Harrogate, also, in a great measure, does the " Champagne " part.

The milder waters from the mild sulphur well, Royal Pump Rooms, and the mild sulphur water from the Royal Baths are less aperient, but at the same time useful in certain cases of dyspepsia, rheumatism, and gout, and are liver stimulants and blood alkalisers; so too is No. 36 new sulphur water, from the Bogfields. The magnesia water from the Bogfields is the strongest in potassium chloride, magnesium carbonate, and is strong in calcium carbonate. It is not so very aperient, but it is one of the best alteratives and blood alkalisers that we possess, and is an antidote, therefore, to gout, acidity of the stomach, and allied ailments.

The Starbeck, Harlow Car, Bilton, and Beckwith

waters, in addition to being sulphuretted, contain also
sodium carbonate, and, being less saline, are most
useful externally as baths, in cases of eczema, where
the discharges are acid. The alkali sodium carbonate
neutralises the acid from the eczematous surfaces.

THE COMBINED ACTION OF THE SALTS IN THE CHALYBEATE GROUP.

I will take first the Ferrous Chloride Waters, viz. :—

THE KISSINGEN—THE CHLORIDE OF IRON.

In the Kissingen water there are aperient as well
as hæmoglobin making elements.

It will be seen that this water is highly charged
with potassium chloride, sodium chloride, calcium
chloride, and magnesium chloride, in addition to the
ferrous carbonate, and barium chloride. In certain
ailments, therefore, it is a most valuable water, and
is most useful in anæmia, where at the same time the
muscular coat of the bowel requires stimulation.
This was a very strong point with the late Sir
Andrew Clark, in the treatment of this ailment in
young chlorotic women. He argued that unless the
bowels were well acted upon, as well as the blood
stored up by iron salts, the result was fæcal anæmia.
In the Harrogate Kissingen water we have the happy
combination thus indicated, which relieves both.

CHLORIDE OF IRON WATER.

The strong feature of this water is the presence of

the ferrous carbonate, which is "over proof," in addition to the ferrous chloride. Barium and calcium chlorides are present, and they both affect the circulatory mechanism in a way, as has already been explained. It is not so aperient, as a whole, as the Kissingen water, but in certain cases of anæmia is masterly.

And while I am upon the chalybeate group, let me say that it is the ferrous, *not the ferric salts*, which are most readily absorbed into the blood. All the Chalybeate wells belong to this class, except the aluminous chalybeate.

The Tewitt, St John's well, Alexandra, pure Harrogate chalybeate, are all useful as tonics, of a mild form, and play their parts, when prescribed in cases for which they are suitable.

Having thus entered fully into the therapeutical action of the bases, the salts themselves, and the combined action of the waters as a whole, I will now proceed to discuss the maladies in which the waters have been proved useful, and the mode of their operation in each case. But before doing this I will make a few

INTRODUCTORY REMARKS ON THE THERA-PEUTICAL, CLINICAL, AND OTHER INDICATIONS.

In the observations which are to follow, I shall endeavour to apply to ailments and their treatment much that I have spoken of in preceding pages. I shall avoid even giving general outlines of treatment, as each case requires its own particular management. It is impossible for any physician, however accomplished, to direct, at a distance, a course of mineral water remedies, and I therefore consider practitioners on the spot, the best able to indicate and carry out the treatment of diseases which may come under their notice.

In selecting cases to be treated by the Harrogate waters, great care is required. Some are apt to imagine that all that is necessary is to swallow so many half pints or pints of the water a day, and they have done everything. This is a fallacy. In some cases, the laxative properties are indicated; in others, the alterative, the hæmoglobin giving element, or all three combined.

I cannot help here remarking that I think many practitioners do not employ sulphur as a remedy nearly as much as its merits deserve. It is a practice or fashion now to look upon old remedies as fossil-like—so too, old practitioners; the young man fresh from college, now-a-days, thinks he can cure every disease by "the new process"—"the old doctors know nothing."

I will give an account of a very instructive case which happened to me as a young man. It was in a case of Tinea Kerion—a bad species of ring-worm, occurring on the head of a young lady with beautiful hair.

At first the spot was small, but it spread rapidly,

till it became a large, boggy, bald place—the size of a five-shilling piece — commonly called "the cart wheel." The usual remedies failed. The mother being naturally anxious, I wrote to a London specialist and used his prescriptions, but the part got no better but worse. It so happened that an old country doctor came to see me on another matter. Many of my younger brethren would not recognise this fine old gentleman on his dappled grey or chestnut cob, his snuff-box, and his dangling seals. Alas! the type has gone out. I asked him to see my patient, and all he said was — "Yes, powder it with sulphur; it was a remedy of my grandfather's." I did so, and from that time the cure began.

MORAL.

"All that glitters is not gold,
Make new friends, but *keep the old.*"

THE THERAPEUTICAL AND CLINICAL PROPERTIES OF THE HARROGATE WATERS.

Before entering upon this subject let me state that I consider these mineral waters, like many others, are most active in the warm months of the year, both for drinking and bathing, say from May till October. Having sent us **suffering**, in many ways, a bountiful Nature provides, quite close, the *healing* springs from mother earth; in the same way as He sends the *healing* balm from the *friendly* dock leaves, growing side by side with the stinging nettle. On the bright May morn, little does the rosy happy maid wot, that in yonder hedge, where she plucks the flowers for her pretty Maypole, she will be stung, and shed bitter tears—soon, however, to be dried by the soothing influence of the cooling herbal unction. In both cases He sends the sting, but He also sends the antidote.

In dealing with diseases which have come under my notice for treatment by these mineral waters, I shall begin with those that affect the alimentary canal.

Stomach Disorders Generally.

When the breath is offensive, the tongue loaded, there are fœtid eructations; the appetite is faulty, general distension of abdomen, and sometimes constipation, sometimes diarrhœa; a course of the stronger and weaker sulphur salines, with appropriate diet, is indicated. Of course, the cause, whether it be errors of eating or of drinking, or of both, must be stopped, but it must not be forgotten that hereditary and constitutional predisposition very often has as much, or more, to do with this condition of things as anything else.

DYSPEPSIA, WITH PALPITATION OF THE HEART, INTERMITTENT PULSE, AND DEPRESSION OF SPIRITS.

Atonic Dyspepsia.—Where there is general debility, want of tone in the coats of the stomach, degeneration of the peptic glands, gastric juice deficient, sensations of weight at the pit of the stomach, also fulness and discomfort after food, with real pain, which is relieved by pressure, where there are feelings of sinking in the epigastrium in a spot "about the size of a shilling," where digestion is protracted and foul gases are generated, where the tongue is large, flabby, moist, and furred, where there is habitual constipation, and the motions are firm, white, and offensive, with absence of bile; where the pulse is weak, and the heart beats rapidly on exertion, where the skin is clammy, and the hands and feet cold; where there is a sense of lassitude, and a feeling of weight on the chest, as of being garrotted, or having a burglar's knees planted on your breast-bone; where the breath is short, with a little cough, and palpitation, I find a judicious course of the strong sulphur water, followed by magnesia and Kissingen, most useful and beneficial.

Another form of dyspepsia, known as *irritative*, gives rise to the following;—burning at the pit of the stomach, increased by food; heartburn, thirst, acidity, vomitings of acid mucus, red tongue, eructations, throat irritable—causing "a sicky" morning cough—sometimes constipation, sometimes diarrhœa; a cutaneous eruption, palms of the hands and soles of the feet burning, pulse frequently feeble, urine scanty and charged with urates. Here, too, the saline sulphur and magnesia waters are most beneficial, and this is easily accounted for when we consider their composition. The late Sir Wm. Gull was especially strong upon this point.

The two conditions described, generally occur in people whose outward appearance is almost the opposite of one another.

The "atonic" sufferer is generally thin—looks as if a hearty dinner would do him good—nervous system "given way," and feels miserable.

His "irritative" friend is generally a robust, ruddy, "comfortable-looking" gentleman, with a sprinkle of the "jolly" look about him.

And yet, for all that, he is wretched, sleepless— often afraid to go to bed for fear of having a bad night, with all the attendant horrors connected therewith. He gets palpitation, and thinks he will be found dead. He feels funny, wave-like sensations about his chest, which are relieved by eructations. He also gets "creepy" feelings about his head, which cause him to stagger, and, of course, he mutters the word—"a stroke."

The meaning of all this lies in the fact that the fermentations of acids that went on in the stomach irritated the filaments of the pneumogastric nerve; these telegraphed, so to speak, to the base of the brain; that branch of the same nerve—the superior laryngeal—"wired" down to the heart, and, being the

ROYAL PUMP ROOM AND OLD SULPHUR WELL.

nerve depressor, inhibited its function; and the result was the chest wave, the intermittent pulse, and the little commotion in the head, causing vertigo.

The explanation of good results in such cases lies in the fact that the sulphur saline waters removed the fæcal matters from the alimentary canal that were the source of irritation.

They also prevented the accumulation of such matters as produce disturbances in other organs— such as head disturbances and the general conglomeration of symptoms alluded to above—affecting the sensory and motor nerves, and the circulation in general ; they likewise neutralised the fermentations of indigestion, and removed the excesses by purgation and diuresis.

CONSTIPATION.

Saline aperients are generally beneficial in chronic constipation. The Harrogate saline waters are particularly so. When there is chronic catarrh, with loss of power in the muscular coats of the stomach and intestines, a *well-regulated* course of the saline water affords marked relief, especially where salts of barium are present.

HÆMORRHOIDS OR PILES.

Treatment by Harrogate salines is often very happy in its results in cases of hæmorrhoids. This is

E

easily explained when we consider the relations of the veins in the abdomen. A smart aperient dose of strong Harrogate saline relieves portal plethora, by abstracting a quantity of serous fluid from the portal blood. Indirectly this relieves the congested hæmorrhoidal plexus of veins around the anus; for the superior hæmorrhoidal vein is a branch of the inferior mesenteric vein, which gathers up, in addition, venous blood by the sigmoid veins, and from the left colic, and, after being strengthened by all these tributaries, empties itself into the splenic vein, which is a branch of the portal.

CONGESTION OF THE PELVIC ORGANS.

This is not the place to enter into all the factors which bring about an over-congested state of the ovaries and uterus. Suffice it to say that the Harrogate sulphur saline waters form an admirable adjunct to treatment in the following states of these parts—namely, in congestion of the ovaries, with the usual pain in the right or left part of the abdomen—low down; with forcing and bearing-down pains generally.

The pathology is congestion of the ovaries and subinvolution of the uterus, which means hypertrophy of the muscular coat of the womb, equally with that of the connective tissue. The cause of this state of things is generally laceration or bruising of the cervix in parturition, the too early getting up after delivery, miscarriages, and child-bearing at a late period of life. I have found a well-regulated course of the internal administration of the Harrogate sulphur saline, and also a continuous course of vaginal douches of the same water, sometimes at a temperature of 110° Fahr., sometimes tepid, and sometimes cold, together with the administration of bromide of potassium and iron, a remedy to be sought after by most who suffer in the manner I have indicated.

The explanation of a course of treatment in these disorders by sulphur saline purgation by the Harrogate waters lies in the fact of their being well borne for a lengthened period, and the fact that branches of the inferior mesenteric vein inosculate with those of the internal iliac, and thus establish a communication between the portal and the general venous system. So it comes to pass that the pelvic congestion is lessened, the ovarian neuralgia done away with, and hypertrophy of the uterus reduced.

CONGESTIONS OF THE KIDNEYS.

These are relieved in three ways, by the skin in perspiration, by the use of purgative salines, and by diuretics, be the complaint acute or chronic.

The Harrogate sulphur saline is diuretic, but in chronic kidney disease this power is assisted by the aperient relief gained by portal derivation, and the explanation is, " that besides the anastomoses between the portal vein and the branches of the vena cava

inferior, the anastomoses between the portal and the
systemic venous system are formed by the communication between the left renal vein and the veins of
the intestines, especially of the colon and duodenum,
and between the superficial branches of the liver and
phrenic veins" (Gray).

Independently, however, of these "communica-
tions" anatomically, there is a much more important
physiological relation between the two excretory
organs (bowels and kidneys), for if Harrogate sulphur salines fail to purge, or purge but in part, they
pass on to the kidneys and act as powerful diuretics.
In dropsy this is most valuable. The magnesium
and calcium salts in the Harrogate waters are
distinctly stimulants of the renal epithelium. When
passing through the cells, these salts carry with them
a quantity of water from the venous plexus around
the tubules, and actually produce diuresis. To this
class of remedies the name "saline diuretics" is
given.

These are chiefly alkaline in their influence on the
blood and urine, but are also independently active
as specific renal stimulants, and in them we have an
indirect means of influencing the venous plexus
around the tubules, and in this manner the whole
renal circulation and the general blood pressure are
relieved.

In congestion of the kidneys I have found the
Russian vapour, or Turkish baths, materially help the
internal administration of the sulphur saline aperient,

by acting on the skin, and both combined will often relieve and cure dropsies from some causes.

CONGESTION OR HYPERÆMIA OF THE LIVER, AND OTHER ENLARGEMENTS.

These may be due to many causes—to wit, exposure to wet and cold; exposure to excessive heat —with malaria; also the enforced use of foul and polluted water for drinking purposes, and errors of diet, such as excesses in eating and drinking. There are also other causes, such as "family livers."

A judicious course of one or other of the sulphur waters, according to the symptoms, is most advantageous in congestion of the liver, or in what is known as "Indian liver."

In these cases the patient has generally suffered from fever, caused by exposure to variations of temperature—sometimes excessive heat, then cold, hard work, and it may be hard living, such as our soldiers experienced in the late Chitral expedition. The liver generally swells, and is tender, and jaundice is often present.

The purgative effect of the strong sulphur water, and a diet properly regulated, relieves the congested portal circulation which lies at the circumference and between the lobules of the liver. At the same time, doubtless some of the salts are absorbed into the blood and excreted by the kidneys, which they powerfully stimulate, and thus open up the urinary discharge, which is the second great channel of relief to the liver. Therefore, I say that the use of the Harrogate saline sulphur water is physiologically and pathologically sound.

Jaundice.—I have referred to jaundice accompanying congestion of the liver, and disappearing with it. The cause is evident—under treatment the pressure upon the bile ducts is removed, and the bile takes its natural course into the intestines, instead of appearing in the skin from being absorbed into the system.

GOUT.

Considerable judgment is necessary in applying the Harrogate waters to cases of gout. If injudiciously taken, especially where acute gout is threatened, they will not infrequently bring on an attack. It is in chronic gout that treatment by this remedy is indicated. It is an old saying that gout is "brewed" in the liver. That is true to a certain extent, but recent research has also more or less pointed out that uric acid is formed from urea in the kidneys. All this is fully gone into by Dr Arthur Luff, in his recent work on gout, only to be contradicted by Dr Chalmers Watson, of Edinburgh, from his recent experiments and researches made there, under similar conditions (see *Brit. Med. Journ.*, 28th Jan. 1899, p. 205). Who, therefore, are we to believe?

It is generally said that an attack of gout is due

to excesses of stimulants and gorging, but family
history, or, shall I call it, "the sins of our forefathers,"
have frequently much more to do with it.
Gout, unfortunately, is in most cases an ailment
which accompanies the patient through all his years
of existence, and be it "suppressed" or "well
developed," periodically has to be dealt with by
remedies and diet. Potash and colchicum, of all
other remedies, "look acute gout straight in the
face," but in the chronic forms the course of mineral
waters from time to time complete that which these
fail to accomplish.

The excesses of urates and uric acid in the blood
are eliminated from the system by a well-directed plan
of Harrogate waters. The strain upon the kidneys
is lessened by the deposits being extracted, and the
general tissues of the body are left in a healthier
condition.

So far, therefore, as the treatment of gout is
concerned, the Harrogate sulphur saline waters
operate freely upon the portal circulation, causing
the presence of bile pigments in the motions, which
means the destruction of hæmoglobin—parts leaving
the body in the fæces by bile, and part by urobilin
by the kidneys. They wash out from the tissues
uric acid, and urate of soda from the tubules of the
kidneys. Indirectly, therefore, gout is relieved and,
in many cases, for the time cured—by the uric acid
being alkalised by the sodium and magnesium and
calcium salts, and by the channel of excesses being

diverted through the bowels and kidneys, as has already been pointed out.

The regular use of saline aperients is especially necessary in gouty persons with contracting kidney and high blood-pressure. How far their utility is to be ascribed to their direct effect in lowering the blood-pressure, and how far to the removal of waste products which might raise the pressure, is "a very nice question," as the lawyers would say. Nevertheless their efficacy is undoubted.

From experiments made with the stronger and weaker sulphur groups—namely, the strong sulphur water, strong Montpellier water, mild Montpellier sulphur, new well No. 36, and the magnesia, with a view to increasing the solubility of sodium biurate (uric acid) in distilled water, it was found, generally speaking, that the strong mineral waters do not increase the solubility of sodium biurate, but the mild Montpellier does, as also the magnesia and No. 36.

How far all this can be applicable to their action on the blood, when charged with sodium biurate, has not been, so far as I know, proved in practice. At the same time it seems to be in accordance with the general results by treatment by the sulphuretted and sulphided waters of Harrogate, and also by sodium *sulphated* mineral waters of other spas, such as Leamington.

I believe, also, the mineral waters have been added to artificial serum, charged with biurate of sodium solutions, and in this way used as a chemical test,

but, personally, I look upon these as "pretty"(?)
laboratory experiments—interesting to the chemist,
but when tested in the labyrinth of human flesh and
blood, in life, as having little practical value.

SCIATICA OR HIP GOUT.

In approaching this ailment I feel that to some
varieties the treatment by the Harrogate sulphur
saline water is not applicable. I also feel that I am
confronted by the opinion of many physicians, who
say that not infrequently sciatica gets well under
no particular treatment.

In fact, from time immemorial, the cause and
treatment of sciatica have been vexed questions.

The local causes of sciatica are, most frequently,
long-continued sitting, colds or draughts upon the
buttock, and sitting upon damp substances; but not
uncommonly the cause is associated with gout or
rheumatism and mental depression. I have found
the Harrogate sulphur salines promote a cure when
all other remedies have failed, and I cannot speak
too highly of the happy results obtained by this
mode of treatment.

This cure, I admit, was materially assisted by
treatment by superheated dry air—so useful in cases
of neuritis.

RHEUMATISM.

The forms of rheumatism which I have found
benefited by a course of the Harrogate sulphur

saline waters are "muscular rheumatism," lumbago, stiffness and pains in joints and muscles after repeated attacks of rheumatic fever, or of general rheumatism; chronic articular rheumatism of those advancing in years, and chronic "rheumatics" generally.

The warm sulphur saline or Turkish baths, also the Aix and Vichy baths, often materially assist the internal administration of the water; and massage, medical rubbing, together with the direct application of the douche to the affected parts, when persevered in, do great good.

The superheated dry air treatment also answers admirably, in conjunction with the waters internally.

I will not weary my readers by reiterating all that I have said by way of explaining the action of the Harrogate sulphur salines in rheumatism. Themselves "alkalisers" of the blood, they act upon the plasma indirectly, by combining with the rheumatic poison, whatever it may be, and carry it with them out of the system by virtue of their diuretic influence. In addition to and in combination with this power, the value of the plasma is affected by their purgative properties, as seen in the influence they exercise upon the salts, water, and other constituents of it in the portal system.

ANÆMIA—CHLOROSIS.

Anæmia means deficiency of the red blood corpuscles, and may be due to their imperfect formation, to direct loss of blood, or excessive destruction.

It is to the last I would direct attention, under the name of chlorosis. Females are generally most subject to this ailment from fifteen to twenty-five years of age.

Chlorosis, or "Green sickness," is met with in all ranks of life, and the chief phenomena are—pale conjunctiva, pale gums, tongue, and lips; subjects stout enough, but with waxy look, and in some cases a kind of pea-green, pasty look, or faces like alabaster, and bloodless.

So great is the whiteness sometimes, that on looking at a patient, I have been almost sensible of the reflection being thrown upon my black coat.

These patients generally suffer from constipation, dyspepsia with eructations, headaches, dizziness, noises in the ears, neuralgia in various parts, especially in the left side, and attended with tenderness, violent palpitations, sleepless, dreamy nights, shortness of breath on going upstairs or walking uphill, derangement of menstruation, and general lassitude.

There are hæmic murmurs at the heart, and the pulse is compressible, feeble, and weak.

Such a state of affairs may lead to tuberculosis if not attended to; and it is with a view to proving the powerful influence for good of the Harrogate saline ferrous water, as found in the Kissingen well, that I have taken the pains to succinctly enumerate a few of the leading manifestations of this disease.

Strong as is my belief in the curative properties of the Harrogate sulphur saline waters in the previously related forms of disease, yet equally—nay, I may say, more strong is it in the ferrous group in cases of chlorosis in young women.

It is most important in treating patients suffering from chlorosis, that the bowels should be kept acting. We have in the Kissingen water that masterly combination of salts, which not only does that, but which is also hæmoglobin making.

The chloride of iron water, too, is quite a specific in anæmia, containing, as it does, large quantities of barium chloride and calcium chloride, and in addition, ferrous chloride and ferrous carbonate, and is therefore most valuable. The action of the barium and calcium chlorides upon the muscles and the heart make it so. All this has been already described.

Not infrequently have I found, on prescribing chalybeates alone, that hæmoglobin would not go beyond a certain point, but by giving a natural saline, in combination with the natural ferrous compound, benefit was made manifest in a short time. Therefore I conclude that for chlorotic young women—or women who are "waxy looking," or for others whose faces are like "alabaster"—with all the constitutional disturbances herein described, one or other, or a combination of several of the ferrous group is almost a certain specific.

Rosy cheeks take the place of pea-green faces, hearts that beat in discord now beat in tune, and all the other dependent troubles vanish one by one, like dew before the morning sun.

In treating cases of anæmia, it is most instructive to watch the blood by the aid of a hæmoglobinometer and hæmacytometer.

I remember a lady, at the end of a fortnight's treatment, herself saying, after I had pricked her finger and put the blood, mixed with water, into the receiver, "That looks much redder." Her observation was proved to be correct.

CLIMACTERIC DISORDERS, OR DISORDERS DURING THE CHANGE OF LIFE.

When there is a headache, especially in the back of the head, when there is aphasia, when there are epileptiform attacks, when there are heats and chills, giddiness or vertigo, when there are disorders of the digestive organs, associated with a tendency to grow fat, when there is derangement of the liver, kidneys, and skin, when there are constipation and flatulent distension of the intestines, when there is lithiasis or passing of gravel or red sand, a proper course of Harrogate saline does great good. The physiological and therapeutical actions of the principal salts, which have been previously dealt with, give the proper explanation, and I can with confidence recommend these waters as being a pronounced assistance to every physician or surgeon who makes the diseases peculiar to women his particular study.

STRUMA.

When the glands in the neck, under the chin, or in other parts of the body become enlarged and suppurate, and discharge cheesy matter, as well as pus; when there is scrofulous disease of joints, "white swelling"; when the alimentary canal suffers from stasis or atony in its whole course, complicated with tabes mesenterica (consumption of bowels)— in fact, when we get the usual scrofulous constitution —a saline course, with ferrous mineral waters, is of great value. The chemical composition of the water and the pathological condition of the parts affected fully warrant such success. The chloride of calcium has already been referred to as having anti-strumous properties, and the various maladies connected therewith have been dealt with jointly and severally under the headings—chloride of calcium, chloride of barium, and chloride and carbonate of iron.

LEAD POISONING.

When the breath is offensive, where there is a blue line on the gums at their junction with the teeth, where the skin is dry and sallow, where there is emaciation, where the pulse is too slow, where there are tremors of the muscles, where there is "drop wrist," the internal use of the saline sulphur water and the baths considerably assist other remedies, such as iodide of potassium and cod-liver oil. Massage and medical shampooings and saline warm douches are all adjuncts to the treatment, especially where there is local paralysis. The Aix and Vichy massage baths, in a wonderful way, help internal administration of the mineral waters.

All cases of lead poisoning require aperients to help the elimination of the lead poison from the system. We have generally obstinate constipation to deal with;—in severe cases, "lead colic," partly arising therefrom.

The strong sulphur aperients, therefore, in their natural state, are most valuable adjuncts; and a prolonged course often brings a happy termination to the above mentioned troubles.

The sulphur, Aix and Vichy massage, and Turkish baths are also useful adjuncts to the treatment internally.

SKIN DISEASES.

ECZEMA.

Eczema, especially when associated with chronic gout, which it often is, demands some notice.

All physicians who have had to deal with this troublesome complaint know what a worry it is to them to cure or even palliate it. And what about the poor sufferer—his nights of suffering, his days of

misery, his never ceasing round of applications, the rags
hot, the rags cold, the soothing (?) lotions, the stinging unctions !!! Oh! what a misery—the rubbings
and scratchings, even to bleeding. "You must
not rub the sores," says the doctor. Oh no, but what
about the night? When he sleeps, he rubs uncon-
sciously, and often the most carefully and artistically
applied bandages fail to keep adapted the healing
balms. And when the raw surface is found bare and
exposed in the morning, the only apology is—" I did
it in my sleep."

I should be sorry to lead such a one to think that
we have " the miracle " in the Harrogate waters, on
the " open sesame " system. Certainly not. There
generally is a cause—and when the cause is removed
the effect will cease. In addition to the administration internally of these waters, we have also the
baths.

The waters of Harlow Car, Beckwith, and Starbeck
are best adapted for bathing purposes in such cases.
And this can be easily understood when we examine
the analysis of their mineral waters. They all
contain little chlorides, and, in addition to sulphur,
they have also a fair amount of alkali as carbonate of
sodium. The acrid discharges of eczema being acid,
are therefore counteracted by the alkaline elements
in the three above-named mineral waters. I have
found also that the superheated dry air sweats the
acid out of the system and dries the irritable sores in
a wonderful way.

PSORIASIS.

Before I came to reside in Harrogate, I had been in the habit of sending patients suffering from this ailment for treatment there. I well remember the case of a young lady, which was as bad as bad could be. I did all I knew. She consulted that eminent specialist Dr Robert Lieving, of London, used his remedies, but all to no use.

I sent her to Harrogate. She was under Mr Hunt, and, to make a long story short, in two months, to all appearance, she was quite cured. The rash was on her knees, elbows, thighs, arms, head. I cannot, however, say that " the cure " was permanent, for nine months after, she had a slight return of symptoms. She again came to Harrogate, took the waters and had the sulphur baths, with the result that now her skin is quite well, and she is going to get married, — a proceeding out of the question till she underwent sulphur treatment two seasons in succession.

Acne, Lichen, Prurigo. — Obstinate cases of acne, lichen, and prurigo, when mixed up with struma or gout, or where the liver and stomach are disordered, yield very often to the combined effects of the different kinds of mineral waters in Harrogate. Regular habits, regular hours, and a regulated diet are also imperative in such cases.

Herpes zoster or Shingles. — Most medical men have found the neuralgia which accompanies this painful

F

complaint, and which in those advancing in years follows it for weeks and months together, a vexed and worrying problem as to treatment. At any rate, I have. And I am glad to be able to record the fact that I have, I feel sure, advanced the treatment in a satisfactory manner, by the internal administration of the saline water and the baths. It is a long, trying, and tedious malady to conquer, especially when the patient is more than sixty years of age, and if anything can be found to assist other remedies one is very thankful. I have found the Harrogate waters successful in such cases.

OBESITY.

To some, perhaps, it may appear odd that I have chosen to class this as an ailment. But yet it is so. The distress which it causes—both bodily and mentally—no one knows except those who suffer from it. It prevents one from getting about, I mean from enjoying walking, cycling, riding, shooting, and all ordinary exercise, which one may have been used to when thinner. Then again, one does not like to be continually remarked upon. "How stout you are getting!" and so on. Apart from that, too, there are domestic and family feelings to be considered. It generally happens that a fat man or woman *snores* at night. If either one or the other happens to be married, and his or her partner in life be a light sleeper, the nights are disturbed, and it means a separate bedroom. This is not always attainable, neither is it always desirable. I have had many tales of woe on this point. It prevents married couples from accepting invitations, and if they go to hotels, it is an extra expense, and they fancy they are remarked upon. These may seem little matters to some, but in the minds of highly sensitive people they are mountains. But some will ask, "What does it matter so long as you are well?" That may be so, but too much fat is a nuisance and becomes a disease. Of course, all this may be carried too far, for from a long experience one finds that the public are never satisfied —if they are fat they wish to be thin, and *vice versa*.

In beginning a course for the reduction of fat, it is most important that the various vital organs of the body should be found to be tolerably sound, especially the kidneys. Grave harm may be done by beginning and carrying out a reducing regimen, unless the urine be frequently examined and found to be non-albuminous. If the fat can be reduced with safety, well and good, for an excess of it only enfeebles the heart, hampers the respiration, and draws from the other organs of the body that nourishment which they so much require.

Diet, of course, is a most important factor. If a man with a tendency to obesity goes in for beer, stout, champagne, sugar, bread, and potatoes, and no daily exercise, of course he lays on fat more and more.

In going in for a course of reducing fat, the patient must make up his or her mind for a struggle. He must go steadily and consistently on and on with the treatment, and put up with hunger and thirst, and take plenty of exercise, especially golf. A course of the various mineral waters of Harrogate, under proper medical supervision, form admirable adjuncts—a good purge from the old sulphur well in the morning, and saline chalybeate restorer to weak and flabby tissues later in the day, are what is wanted. Turkish baths are also most useful.

I give now a typical case of obesity, in which 42 pounds in weight were taken off with safety and comfort from a patient by the mineral waters and diet.

A strong gentleman aged 50, muscular and ruddy, began the treatment on July 25th. His vital organs were all tolerably sound and performing their functions satisfactorily, except for the distress arising from the presence of excessive fat. From circumstances connected with his calling, he was obliged for years previous to lead a sedentary life, and because he worked hard with his brain, he lived well.

I put him upon the following regimen and diet:

Before breakfast, 24 ounces of strong mineral waters. Breakfast, a little fish, one egg and a cup of tea—no bread, butter, milk, or meat. Played golf for 1½ or 2 hours in the morning.

Luncheon—a chop or two, or a big slice of mutton or any meat, green vegetables, salad, little cheese and one biscuit, about a glass of whisky well diluted with water. Rested for two hours and read papers, etc. Walked for one or two hours in afternoon. Took

10 ounces of either chloride of iron or Kissingen water at 4 p.m.

Tea—one cup, infused three minutes. Nothing to eat. Had a Turkish bath three times a week.

Dinner—fish, meat of any kind, green vegetables, salad, little cheese and one biscuit—one or one and a half glasses of whisky with water. Played billiards after for an hour or more and went to bed early.

He was 17st. 10lbs. on July 25th, on October 31st he was 14st. 10lbs.

This treatment was carried out under great care, and constant watching, with the result that now, January 31st, he is able to hunt, do as he likes, and is in all respects a better and a happier man.

The human system bears a long continued course of the aperient waters from the old sulphur well and we have, next door to it, the restorative chalybeate waters to build up the enfeebled muscles, especially the heart muscle.

I therefore conclude that where there is obesity, general distress, shortness of breath, gouty tendencies, and fatty infiltration of the muscles, especially affecting the heart, a well-regulated course of combined Harrogate waters helps the crippled frame from day to day, gives tone to the system and prolongs life.

THE BATHS.

At the Victoria Baths and at the Royal Baths there is a splendid system of bathing.

Nothing that money and balneological experience can produce have been spared to bring this about. The Turkish bath is perfect, so are its attendants.

There are in addition the "Aix douche" or "douche-massage," and the Vichy massage baths; the Harrogate combination, local vapour and general Russian vapour baths, needle baths, sulphur baths, and inhalation and pulverisation rooms, etc.

All these baths are useful in muscular and various other forms of rheumatism, in gout, derangements of the liver and stomach, in diseases of the nervous system, and in some diseases of the skin, such as eczema, psoriasis, and acne.

The hot internal douches are most beneficial in complaints from which women suffer, especially after child-bearing.

EXAMINATION OF THE BLOOD WITH REFERENCE TO THE APPLICATION OF HARROGATE MINERAL WATERS.

There is a well-known saying, with reference to remedying all errors, "remove the cause and the effect will cease."

In many of the diseases which find relief by Harrogate mineral waters, the cause is shown to be in THE BLOOD.

This has led me to examine the blood of all patients suffering from gout, rheumatism, anæmia, and allied ailments. For an elaborate description of the blood,

I refer the reader to a *Text-Book of Physiology*, edited by Professor Schäfer.

Suffice it to say here, that there are various methods of determining the percentage of hæmoglobin in the blood, and also the numerical corpuscles —that of Gowers, that of F. Hoppe-Seyler: that of v. Fleischl, and Oliver's improvements on v. Fleischl as adapted by Lovibond, of Tintometer fame, at Salisbury.

To be brief—

(1) The word *normal*, as applied to the blood, is used to indicate the average condition of the hæmoglobin and corpuscles found by experiment (Oliver) in the blood of a large number of healthy persons. This point was made the 100 on the scale. More extended observations might show that the present 100 is either higher or lower than the average.

100° on the scale corresponds to 15·5 of hæmoglobin as determined by chemical analysis.

(2) On the hæmoglobinometer, the 100 corresponds to 4000 corpuscles per cubic centimetre.

(3) The blood ratio is obtained by dividing the hæmoglobin number by the number for the corpuscles—thus

(4) $\dfrac{\text{Hæmoglobin.}}{\text{Corpuscles.}}$

or say $\frac{100}{100} = 1$ in the case of normal blood.

Nevertheless $\frac{80}{80}$ may be normal quite as much as $\frac{100}{100}$, the quantity of corpuscular matter varying from individual to individual, being higher in man than in woman, and depending apparently on the size of the capillaries. When hæmoglobin is higher than corpuscles, it indicates a gouty and apoplectic tendency; when the reverse is the case, anæmia is indicated.

It is in such cases where the treatment by Harrogate mineral waters is attended by happy results. I have already stated that destruction of hæmoglobin is achieved in the liver, and may be assisted by administering mineral waters, such as the strong sulphur, which acts on the liver and portal circulation. Therefore, having examined the blood by the hæmoglobinometer, and finding it too rich in hæmoglobin (gout and apoplexy), naturally a remedy which assists the liver as a hæmoglobin destroyer is indicated.

Then, again, when hæmoglobin is wanting after examining the blood by the hæmacytometer and hæmoglobinometer, we know that iron is required, and that element we have in the various iron wells.

Of course, in a work like this, all the variations during the different parts of the day cannot be gone into, nor can the position of the body, nor the effects of altitude, be discussed.

On the whole, however, I am decidedly of opinion that the hæmoglobinometer and hæmacytometer are gains in diagnosis, and also in indicating lines of treatment, by the different Harrogate mineral waters.

Treatment by Superheated Hot Air in Rheumatoid Arthritis, Gout, Rheumatism, etc.

For some months previous to April 1898, I had been watching the reports of cases by this mode of

treatment, and as I had several suitable ones. I
resolved to put it to the test. There was no public
appliance in Harrogate at the time, so I hired one for
my private use.

There are now several kinds of apparatus invented
—Tallerman's, Dowsing's, and Grevill's. Mr Tallerman is the inventor of the system, and his apparatus
was at first heated by gas. It can now be done by
gas or electricity. Dowsing and Grevill use electricity only.

On 10th May I began the treatment. It was
carried out by a trained nurse, specially instructed in
the work. The cylinder into which the affected
limb is put is protected by asbestos lining, is heated
from below (when done by Tallerman's method), and
is open at both ends. At one end is a lid, which
can be screwed on and off as desired; the other end
is encircled with a piece of Mackintosh sheeting,
which is drawn over the limb and tied. The limb,
before being placed in the cylinder, is wrapped in
one or more folds of lint; the temperature being
previously raised to 180° Fahr. in the apparatus.

The heat is now gradually raised, and continues
to go up to 260° Fahr., and even 300° Fahr. I have
seen it asserted that patients can stand a temperature
of 400° Fahr., and even more. This I do not believe.
The chamber may register that, but the limb itself
never could stand that heat. The only explanation
is that the quantity of lint must be such as to make
a difference of, at least, 150° Fahr. less. In an oven

at 400° Fahr. a pie will bake. It is a pity, therefore, that exaggerated statements are published, which only tend to throw discredit on this treatment. The limb remains in the cylinder for forty-five minutes.

The effect of this heat is to make the patient perspire freely all over, and generally to raise the body temperature one, two, or even three degrees.

My first case was that of a lady suffering from rheumatoid arthritis of some five years' standing. She had tried all the usual treatment of baths and waters at Harrogate and Leamington, off and on, for that period, but got no benefit, but became worse.

Most of her joints were affected, especially the wrists, knees, and ankles. So crippled was she that she could not walk downstairs, but had to manage by sitting first on one step and then on another, and wriggle herself in that way by the aid of her arms; and when she took the railway journey here, she had to come in her bath-chair in the luggage van, as she could not be lifted into the carriage. I shall never forget the difficulty we had to get her worst leg into the cylinder. She was weak and wasted and especially nervous. The heart's action was feeble, and the pulse 68. Temperature, 97° Fahr. Her leg was in the cylinder forty-five minutes. The effect was free perspiration all over, that she was at once easier, her body temperature 99°·4 Fahr., pulse 100, regular and full.

The temperature in the cylinder rose to 254° Fahr.

This treatment went on every other day, and every day with an increased temperature of cylinder to 300° Fahr., till 27th June, when she had had in all thirty-two applications.

To make a long story short, by the end of the first month she could walk with a crutch to church, and when she went home she could get about nicely, take short walks, and go up and down stairs. She gained 14 lbs. in weight. The internal treatment was a course of sulphur water, tonics, and cod-liver oil. From time to time she came by train and had an application, and in the course of October she had twenty-one. The improvement has continued, and now, March 1899, though not exactly cured, she is in every respect wonderfully better, free from pain, and can get about.

My next case of a similar kind was that of a gentleman, sent by a physician from Bradford, age 41.

He was suffering from rheumatoid arthritis, more or less in all his joints; they were all swollen, and his knees were fixed—his legs at right angles to his thighs. He had been bedridden almost for ten months, could not lift his feet from the bed (they had to be moved for him), could not put his hand to his head, and was in dire pain all over him.

He was greatly reduced in flesh, and was sick daily.

He was too ill to come by train, and had to be moved from Bradford to Harrogate in an ambulance carriage.

He had his first application on 6th June, and to tell the truth I was almost afraid to begin this treatment. However, I did.

Temperature of body before, 100°·2 Fahr.; after, 101°·6 Fahr. Pulse, before, 100; after, 100. Temperature of cylinder, 260° Fahr.

The treatment was continued every day at first, and every other day later on, till 3rd October, when he had had sixty-two applications, and he went home to Bradford quite well and able to get about and walk a couple of miles.

Now this is a most remarkable case. Here was a man going from bad to worse, made well, and would, I am sure, have died within a few months.

He had been well and skilfully treated by his doctors at home, and medicine failed where the hot air wrought the cure. His diet during the whole time was milk only, and he got quite fat before he left. On 23rd December the patient writes :—" You will be glad to hear that the benefit derived from your treatment has, so far, been permanent."

Anxious to find out how it was that my patients were deriving benefit, I examined the blood before the treatment and immediately afterwards, and had some most interesting results. I am not going into this at length now, but speaking generally, if by the hæmoglobinometer the hæmoglobin was, say, 102 before the treatment (or bath, we will call it), after, it would be 90; and if the corpuscles, by the hæmacytometer, were 90, they would be 94 after.   How, then, came the destruction of hæmoglobin ?

So far as present knowledge goes, the destruction of red corpuscles takes place entirely in the liver, producing bilirubin, hæmato-porphyrin, and other colouring matters, no destruction taking place in the blood-vessels.

The improvement effected by the superheated hot air is, therefore, probably almost entirely due to the dilatation of blood-vessels allowing greater proportion of blood to flow past the diseased portions, and consequently greater chemical action.   This would allow the alkali in the blood greater opportunities of dissolving the deposited urates, and the high temperature would assist in rendering the salts more soluble.

At the same time, the greater activity of the glands of the skin under the influence of the hot, dry air would tend to remove the urates from the blood by secretion as fast as they were taken up by the morbid tissues.

During " the season " I treated seven more cases of very bad rheumatoid arthritis, all with marked benefit except two, and in them there was neither benefit nor harm done.

The conclusions which I have come to with regard to this treatment in rheumatoid arthritis are these:—

1st. To get benefit, the patient must not be too old.

2nd. The complaint must not be over, say, ten years' standing.

I treated several cases of gouty neuritis with marked benefit and even cure, when other treatment, such as sulphur, Aix and Vichy baths, had failed to give relief.

Stiff joints from chronic gout and rheumatism yielded to treatment in a wonderful way, certainly, as well as by the usual baths in use at Harrogate.

During "the season" my trained nurse administered the treatment 330 times to different patients —in rheumatoid arthritis, gout, rheumatism, eczema, anaemia, sprains acute and chronic, and in one case of gout and glycosuria with enlarged liver.

This last case is worth recording. It was that of an old gentleman of 68 years. He had been to Harrogate the year before, had taken the waters, and had used the baths.

He was in a very weak and shaky condition, took very little food except soup, drank whisky and water, and was sick daily.

He was crammed full of gout at every point, and could walk with difficulty on account of pain and stiffness.

I suggested the baths again, but he said they did him no good. I also proposed the Turkish bath, but he said he could not breathe in it, and that it made him feel faint.

I therefore tried hot air, and he had his first application on 25th May.

Temperature of body, 97°·2 Fahr. before.

„   „   99° Fahr. after.

Pulse before, 48 ; after, 60.

Temperature of cylinder, 250° Fahr.

In all he had eleven " treatments," and left for the North on 20th June, able to walk a mile; and a month after, his doctor at home reported that " he was better than he had been for five years."

With reference to this case and several others of the very severe ones, it was necessary for me to watch the patients most carefully while under treatment. I was obliged once to administer stimulants and to discontinue the treatment. And this brings me to refer to statements made by people who have not had experience to the effect that " an ordinary attendant can apply the treatment, without the patient being watched by a doctor, with perfect safety." This may be so in ordinary cases who have very little the matter with them, but I contend that, in severe ones such as I have described, the doctor ought to be in and out for the first few " treatments " at least, to watch the case and to make it safe. I contend that treatment which raises the temperature two or three degrees is no ordinary experiment.

What would people think of a medical man who would leave an ice-pack in the hands of a nurse in a case of fever to lower the temperature from 110° Fahr. to 105° Fahr., and not be present himself ?

The same reasoning would apply to such severe cases as I have enumerated. But this is where the abuse of the hot-air treatment will come in. It will, I fear, be used as a commercial undertaking, and then its merits will not be properly appreciated, but abused. I made a point of being almost invariably present at the whole of the 330 "baths," and I am persuaded that my patients had more benefit by my doing so. I do not say that I was present for the forty-five minutes at each "bath," but I was in and out all the time. Almost in all the cases treated, the mineral waters of Harrogate were given internally, and I believe the dry, bracing air did much towards helping the treatment, from day to day.

Finally, I quote from *The Lancet* of 7th May 1898 : —" We feel sure that before long the method of applying air at hot temperature to diseased portions of the body will take the place it deserves in the estimation of medical men as the *most satisfactory method* at their disposal of *treating* many hitherto very intractable morbid conditions."

But let me add—let it be only used in suitable cases; let it be properly used, and not *abused*.

CONCLUSION.

And now comes the end. I am sorry, in many ways, to leave my subject; but even now I fear I have, like the silkworm, "spun my thread," it may be, too long. No matter what remedies, new and old, we as physicians apply, the END MUST COME, and yonder churchyard will be our final resting-place, with brighter HOPES for another Country, where

sulphur is not needed to purify, nor iron to strengthen.

Having done our best, we must leave the rest. If I have wandered from my text at times, and have introduced illustrations, perhaps, not always medical, the reader will excuse a mind which, retentive to a degree, sometimes *boils over*, but which never, I hope, forgets the sympathetic side of human nature.

And with this apology I bid the reader adieu.

[APPENDIX.

G

APPENDIX.

COLOUR CURVES.

THROUGH the kindness of Mr J. W. Lovibond, I am able to give the Tintometer colour curves of some of the Harrogate waters analysed by Mr Townsend, as well as the curve of the Moffat sulphur water, the Chalybeate water of Strathpeffer, and of Challés or Lady Cromartie's Well, which is a sulphur water from the same locality.

In many substances it has been found that the variations in the colour curve gives an unfailing indication of an alteration in the condition of the material, and there is a possibility that this principle may be applied to water analysis.

It will be noticed, by inspection of the curves, that there is a great similarity between the curves of the strong sulphur water (Harrogate) and the Moffat sulphur water, the yellow curve and the black being very characteristic. The curve of Well No. 36 is of quite a different character, although the colours are the same.

The curve of the last-named water somewhat resembles that of the Magnesia water. The curve of the Chalybeate or Saints Well resembles that of the Kissengen Spa water, and not that of the Chloride of Iron water.

The Tintometer is in regular use at the laboratory of the London County Council, the Liverpool Corporation at Lake Vyrnwy, the Massachusetts Board of Health, etc.; and it is hoped that by the collection and publication of data relating to the colours of the waters, we shall, at no distant date, be enabled to draw more certain conclusions as to their chemical nature.

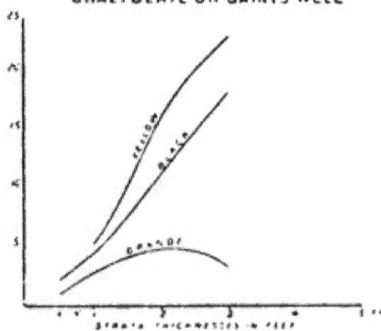

WELL Nº 36

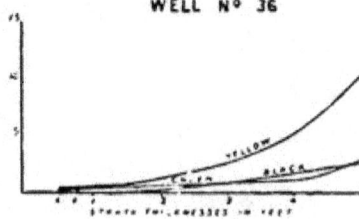

MOFFAT SULPHUR WATER

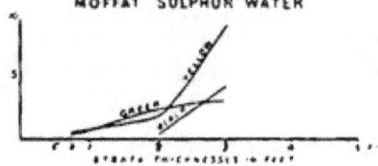

CHALYBEATE OR SAINTS WELL

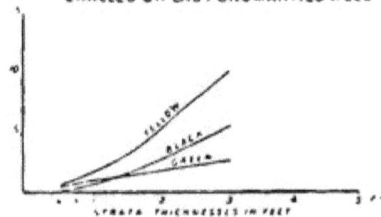

CHALLES OR LADY CROMARTIES WELL

STRONG SULPHUR WATER

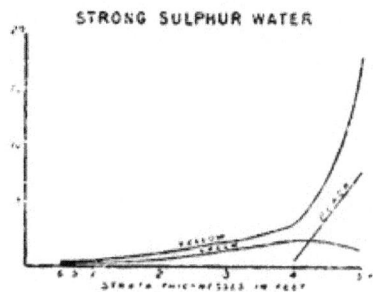

STRATA THICKNESSES IN FEET

MAGNESIA

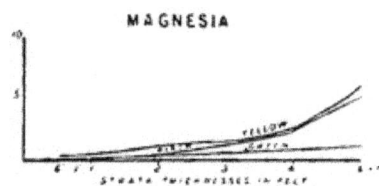

STRATA THICKNESSES IN FEET

CHLORIDE OF IRON WATER

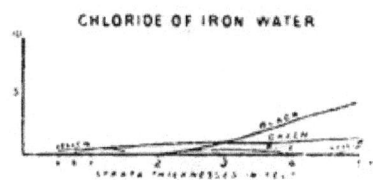

STRATA THICKNESSES IN FEET

KISSINGEN WATER

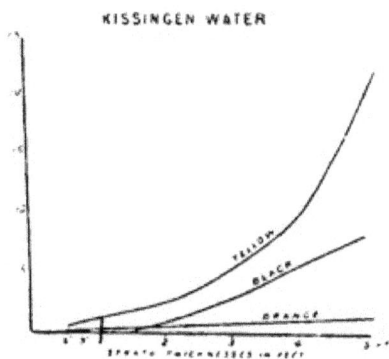

STRATA THICKNESSES IN FEET

Demy 8vo.   Cloth. 2s. 6d. net. *Postage 3d.*

Practical

THE
SECOND
EDITION.

Radiography.

Entirely
Re-written.

*A Handbook of the Applications of the
X-Rays, with many Illustrations.*

BY

A. W. ISENTHAL
AND
H. SNOWDEN WARD

(Editor of " The Photogram").

❦ ❦ CONTENTS. ❦ ❦

LONDON:

DAWBARN & WARD, Ltd., 6 Farringdon Avenue, E.C.

COMPLETE LIST OF BOOKS POST FREE UPON APPLICATION.